TELECOMMUNICATIONS
CONVERGENCE

STEVEN SHEPARD

TELECOMMUNICATIONS CONVERGENCE

HOW TO PROFIT FROM THE CONVERGENCE OF TECHNOLOGIES, SERVICES, AND COMPANIES

STEVEN SHEPARD

McGraw-Hill
New York   San Francisco   Washington, D.C.
Auckland   Bogotá   Caracas   Lisbon   London   Madrid
Mexico City   Milan   Montreal   New Delhi   San Juan
Singapore   Sydney   Tokyo   Toronto

Cataloging-in-Publication Data is on file with the Library of Congress

McGraw-Hill

A Division of The McGraw-Hill Companies

2 3 4 5 6 7 8 9 0 DOC/DOC 0 9 8 7 6 5 4 3 2 1 0

ISBN 0-07-136107-3

The sponsoring editor for this book was Steven Chapman, the editing supervisor was Steven Melvin, and the production supervisor was Sherri Souffrance. It was set in Fairfield by D&G Limited, LLC.

R. R. Donnelley & Sons Company was printer and binder.

McGraw-Hill books are available at special quantity discounts to use as premiums and sales promotions, or for use in corporate training programs. For more information, please write to the Director of Special Sales, McGraw-Hill, 2 Penn Plaza, New York, NY 10121-2298. Or contact your local bookstore.

 This book is printed on recycled, acid-free paper containing a minimum of 50 percent recycled de-inked fiber.

For Bina.

Thank you.

CONTENTS

ACKNOWLEDGMENTS

For professional assistance with the content of the manuscript, I wish to thank the following people without whose technical expertise the book would be much shorter and far less appealing: Gary Kessler of Symquest; Gary Martin of TM2 Inc.; Bob Dean of Ernst & Young; Phil Cashia, Walt Elser, and Bob Brown of the USC Center for Telecommunications Management; Jack Garrett of IFM Technologies; Barbara Jorge, Miguel Montañez, Carla Krebs, Marta Ramirez, Osmán Chacón, Robert Ruiz, Ted Lipsky, Tony Sicle, Jemima Bowden, Martha Bradley, Mauricio Lopez Calderón, and Idsa Genie of Lucent Technologies; Kirk Shamberger, Mitch Moore, Mike Lawler, Ken Rosenblad, Karen Andrésen, and Walter Goralski of Hill Associates, as well as all other members of the staff whose advice and comradeship helped me greatly.

For help with the manuscript itself, I am indebted to Gary Kessler, Kenn Sato, and my wife Sabine, the finest (and pickiest) editor I could be blessed with. I am also grateful to Steve Chapman of McGraw-Hill, who saw the value in this story and took a chance on sharing it with the masses.

As always, I am indebted to my parents for the vision, and my wife and children for the love and patience. Because of you, I can do what I do. Thank you all.

FOREWORD

In 1999, I spent almost all of my time focusing my professional efforts on the ephemeral phenomenon known as convergence. As a professional telecommunications educator and writer, I became intrigued by the tremendous amount of attention that the "all things over IP" contingent was enjoying. As I studied the various facets of convergence, I found that although there was a great deal of confusion among the telecommunications market players about IP convergence and where the industry was headed technologically, there was at least as much among the customers who rely on the telecommunications industry to help them achieve strategic corporate goals through the deployment of evolving communications solutions. As an educator, I realized that I was equally confused and needed to better understand the phenomenon in order to explain it.

To get my arms around what was actually happening, I conducted a series of interviews over the course of a year with more than a thousand people, both within and outside of the telecommunications industry. I interviewed a wide range of telecommunications professionals, from high-level executives to communications technicians in incumbent local and long-distance providers, competitive carriers, Internet service providers, cable television companies, regulatory agencies, telecommunications consultancies, and academic institutions. Simultaneously, I conducted extensive, in-depth discussions with telecommunications customers from within a large number of vertical industries including financial, health care, government, education, manufacturing, professional services, small businesses, and multinationals.

What emerged from these conversations was fascinating. I came to understand that the term "convergence" has become commonplace today and is generally used to reference that inexorable technological drift that seems to be underway toward a common network transport fabric such as the Internet or pure IP protocol. Many believe that today all roads lead to the Internet, and that it is on the verge of becoming the only network that matters. In fact, if we are to believe some industry journals and the word of trade show proselytizers, a week from this Thursday the forklifts will arrive to "upgrade" the circuit switches in telephone company central offices. This upgrade will be accomplished by replacing them with low-cost, universal routers that will do it all—voice, video, images, data, and high-quality audio, all transported simultaneously, effortlessly, and with perfect quality over a common network fabric. Needless to say, although this dream may transpire sometime in the future, it is not imminent. Furthermore, although the technology is centrally important, convergence is much more than a technological event.

In reality, convergence comprises three parts that involve not only technological convergence, but the convergence of companies and services as well. In fact, technology, although important, is only a piece of the process. As Claudio Monasterio of Lucent Technologies in Santiago, Chile observed in a conversation one morning in Mexico City, *"La tecnología es nada mas que un detalle"* (the technology is nothing more than a detail).

The relationship between the three components of convergence is crucial to an understanding of how the phenomenon affects the way corporations jockey for competitive positions in a market where the rules of engagement change routinely and where the players evolve, amoeba-like, alternately splitting into new corporate organisms or merging, enveloping each other with organizational pseudopodia. The pace of doing business is accelerating at an almost immeasurable rate, and although the old adage, "Time is money" still applies, a variation on that theme, "Speed is profit," is perhaps more appropriate in today's telecommunications market. The law of primacy holds true here, and the first to market wins.

The relationship between technology, companies, and services is a simple one. In spite of the technological bias that exists among many players in the telecommunications game, the ultimate product that customers want to buy today is not technology, but rather the services that telecommunications technologies make possible. The technology becomes nothing more than a means to an end, a facilitator. Furthermore, as telecommunications companies take part in the acquisition "feeding frenzies" that are so characteristic of the industry today, they merge their collective financial and intellectual assets to create new companies with pooled technology resources that translate into full-service solutions that deliver enhanced, competitive advantages. In reality, the process is circular. Companies in the telecommunications industry, faced with growing, fierce competition (often for the first time), recognize with chagrin that they do not have the wherewithal to go it alone. To deliver competitive products and services, they must join forces with other players to round out their offered set, a classical convergence of corporations. The "corporate molecule" that results is able to deliver a set of converged technologies, which in turn enriches the product offering.

The technology, then, becomes a facilitator; corporate convergence, an efficiency mechanism; and service, in the form of customer solutions, the desired product. All are necessary, but the first two make the last possible. If the last cannot be achieved, the first two are meaningless.

THE TECHNOLOGY

From a technological point of view, the watchword for the day seems to be "IP." And although IP is most certainly one of the most crucial technologies in evidence, others will play roles at least as central. These include *Asynchronous Transfer Mode* (ATM), which, because of its capability to deliver diverse *quality of service* (QoS) guarantees, will unquestionably provide the core network transport capability in the near term; frame relay, which still enjoys popularity as a private-line replacement technology for a growing range of services, including voice and video; and, of course, circuit switching, which in spite of all words to the contrary will continue to be a central delivery method for years to come. Mike Turner, VP of corporate planning for SBC, recently observed that 98 percent of the traffic carried across the SBC network is data, yet that data traffic contributes minimally to the corporation's revenues. The bulk of them still come from voice.

As IP matures, it will become a more central technology, but its capability to do so will be predicated on both its capability to work well with other technologies to provide granular QoS controls to the customer base and the ease with which it will integrate QoS protocols. These include such emergent standards as DiffServ, IntServ/*Resource Reservation Setup Protocol* (RSVP), 802.1Q/p, *Multiprotocol Label Switching* (MPLS), H.323, the *Media Gateway Control Protocol* (MGCP) suite, and others.

Customers do not want to buy technology. They want to buy services, applications, solutions, and options. They will buy whatever it takes to gain the competitive advantage.

Equally important are the access technologies that customers will rely on to reach the network cloud. These include cable modems, *digital subscriber lines* (DSLs), wireless local loop technologies such as the *Multichannel Multipoint Distribution System* (MMDS) and the *Local Multipoint Distribution System* (LMDS), and more traditional offerings such as *Integrated Services Digital Network* (ISDN) and 56K modems.

We must also consider the contributions of *Synchronous Optical Networks* (SONETs) and *Synchronous Digital Hierarchies* (SDHs), as well as *Wavelength Division Multiplexing* (WDM), *Dense Wavelength Division Multiplexing* (DWDM), and *Ultra-Dense Wavelength Division Multiplexing* (UDWDM) technologies, which stand ready to fundamentally change the economics that drive the telecommunications marketplace. Consider the fact that the central players in the industry—the *Incumbent Local Exchange Carriers* (ILECs), the *Competitive Local Exchange*

Carriers (CLECs), the *Interexchange Carriers* (IXCs)—earn revenues primarily by selling network services, including bandwidth. As optical technologies, such as those mentioned previously, continue to make bandwidth available at ever lower prices and in quantity, simple supply and demand observes that the price will approach zero as bandwidth becomes a commodity. Spot markets for bandwidth are already being created, and companies like Qwest, Level3, and others are billing themselves as "the carriers' carriers" because they can produce bandwidth cheaper than the IXCs due to their ubiquitous and over-engineered optical backbones. For those companies for whom bandwidth represents their corporate bread and butter, what does the dropping price, and hence commoditization, bode?

THE SERVICES

The term "unified messaging" has become common today and represents something of a change in the services world. As networks converge toward a common transport fabric, the power of convergence makes it possible to build applications that can take advantage of that fabric in ways that were previously not possible. For example, some online *Applications Services Providers* (ASPs) now offer free (public domain) global calendar and scheduling services. Customers can register to use the service and can then populate the application with their calendar entries, making the information available to anyone who wants to examine it. Additionally, the application makes it possible to password-protect the information on a client-by-client basis so that the user can create a closed user group by distributing the password only to those individuals that he or she wants to have access to the information. Additionally, the application is collaborative. Once the user group is created, anytime a change is made on one person's calendar, it appears on everyone else's in the group so that conflicts can be detected and resolved. Consider the power of this capability in this world of virtual work groups and collaborative teaming. Commercial applications such as Lotus Notes offer similar capabilities, for a price.

In call center environments that thrive on the ability to quickly, accurately, and repeatedly route information, the ability to direct calls, e-mail, and fax traffic to a specific operator is a great advantage. The concept of unified messaging makes this possible, and with the arrival of IP voice and perhaps even video, online shoppers will have the ability to connect with an agent in a call center in real time to ask a question before purchasing, thus improving the customer service experience and the ability of the agent to collect information about the client.

THE COMPANIES

In the evolving network-based economy, competition will not exist between individual companies, but rather between clusters of companies that together provide a complete solution to a customer's business problem. There has been a feeding frenzy underway in telecommunications during the last few years as key players jockey for competitive positions. Cisco, for example, has purchased more than 40 companies since 1993, providing them with capabilities that are complementary to their data orientation. These include manufacturers of ATM equipment, network management software, circuit-switched and IP voice hardware and software, and *Signaling System* 7 (SS7) signaling equipment, and there is no end in sight to their accretion plans. Similarly, Lucent Technologies, a name long known as a premier provider of circuit-switched telephony products, has augmented its own corporate molecule through the addition of such companies as Ascend (ATM switches), Yurie Systems (ATM access equipment), Livingston (access equipment), LANNET (ATM switching), Prominet (gigabit Ethernet technology) and Octel (voice messaging and response units), to name a few. All the players in the telecom game are buying capability rather than building it, because of two principal observations: (1) In this rollicking competitive market, it is neither prudent nor cost-effective to take the time to build capability in-house. (2) The quickest way to become a leader is to find a parade and get in front of it.

THE CHANGING CORPORATE MODEL

Two forms of company convergence are underway. *Heterogeneous convergence* includes companies that are joining forces with others in a different business as a way of expanding the breadth of their service offerings. *Homogeneous convergence* includes companies that are "acquiring in their own image" as a way to expand their footprint or geographic presence. Both are real, but it remains to be seen which is the most effective strategy.

The changing corporate model is bringing about a shift in managerial focus as well. Traditionally, companies are managed in a top-down fashion. All decisions regarding business activities and transaction flows are made at some central managerial node. This model stems from the industrial age when manufacturing was the money maker and value was created through mass production and driven by the supplier of whatever resources were required to carry out the manufacturing process. All knowledge is concentrated at the top node as well. Lower managerial levels have minimal decision-making power.

Today, many companies have evolved away from hierarchical behavior in favor of a more distributed, customer-driven model where service is the key tenet of success and knowledge the single most important competitive resource available to the corporation. The modern, knowledge-driven enterprise does not compete one on one with other corporations. Instead, it becomes a member of a virtual community made up of consumers, retailers, manufacturers, suppliers, distributors, and even network service providers, and as a community, it competes with other communities. Part of the evolution is the evolution of the network, inasmuch as community members must be able to communicate with one another in a flat, instantaneous, effective fashion. The entire value chain changes, as does the supply chain. It is no longer linear; it has become a "hyperchain."

This evolution is not a painless one. As companies downsize, rightsize, RIF, MIP, and do whatever else they do to achieve critical competitive mass, they discover that the tried and true managerial models of yore no longer work. Successful companies of the '50s and '60s had enormous human resource potential available to them, and problems were best solved by throwing bodies at them in large numbers. They were built on a military model in which individuals were given a single, narrow responsibility and instructed to become extremely good at that single task. As a result, they did but did not develop a sense of the overall mission of the corporation. They were, in effect, managerial drones, working in an environment that in many ways resembled the assembly-line process of a manufacturing floor.

That approach is no longer feasible. Whereas bodies may have represented the competitive advantage flouted by corporations of earlier eras, today the same advantage is achieved through the collection of, access to, and use of corporate (virtual community) knowledge. Individuals are empowered at low levels to make decisions, because it is assumed that they have the ability to collect and act upon information about their customers that will help them make informed, targeted decisions. The term "data mining" is more than just a passing buzz phrase. It is a critical component in the managerial arsenal of competitive corporations today.

Of course, a corporation's ability to ensure that the right information is in the hands of the right people at the right time is solely dependent upon the quality and accessibility of the enterprise network. As corporations evolve to become knowledge companies, networks and the computers they interconnect evolve to become knowledge transport schemes. The degree to which corporations and the virtual communities in which they operate rely on these networks is so great that in many cases the network is perceived to be the single most critical infrastructure component that these corporations have. The numbers support the facts. More than a trillion dollars per year are spent in the United States on information technology, which represents roughly one-sixth of the *gross domestic product* (GDP).

Corporations rely on the network to achieve the pronouncements of their business plans and trust it implicitly. There is, however, a caveat. Today, networks are considered to be "good" if they offer massive quantities of bandwidth to the consumer. However, according to Francis Fukuyama, author of *Trust: The*

Social Virtues and the Creation of Prosperity, "Trust cannot be built from bandwidth alone." Networks today must provide a "technocopia" of value-added services that support a company's drive to become a knowledge-based entity and that go well beyond bandwidth. Given the fact that many of today's corporations seem to be looking to the Internet to provide this capability, some serious questions arise.

WHAT OF THE INTERNET'S ROLE?

Is the Internet up to the task of becoming the network of networks? Consider what it must offer. In addition to providing "unlimited" bandwidth at all levels—in the loop and the cloud—the network must offer minimal delay and jitter, zero information loss, functional reliability, and unquestionable security. Today, the Internet is accomplishing some of these tasks with varying degrees of success. *Virtual Private Networks* (VPNs), often based upon the Internet for wide area transport, have become the most talked about (and perhaps most implemented) technology solution today, and for good reason. They offer the safety, security, and performance of a private network over the inexpensive public Internet infrastructure. They do so, however, with a caveat. They must have some form of added protocol overhead that overcomes the inherent shortcomings of the Internet.

In Steven Spielberg's blockbuster movie, *Jurassic Park,* the chaos theoretician, played by Jeff Goldblum, utters a comment that captures the essence of today's concern over whether all roads lead to the Internet. In response to the park owner's observation that Goldblum isn't suitably impressed by the dinosaur cloning efforts of the park's scientists, he says, "You spent so much time proving that you *could,* you forgot to ask whether you *should.*" Can we transport converged voice, data, and video over the Internet today? Absolutely. Can we do it with any level of guaranteed QoS? Absolutely not. Internet telephony is certainly a viable way to communicate, but in a world that has had nearly 125 years to become accustomed to "toll quality" voice, the Internet does not currently provide an acceptable option.

A QUESTION OF QUALITY

Of course, the Internet is only a piece of the overall solution. In reality, the most effective service providers (and we must accept the fact that the definition of a service provider now changes to go well beyond the traditional access and transport company) will sell a *cloud*—that is, their "product" will be access to a transport mechanism capable of handling all manner of services and able to deliver every possible and desirable QoS level according to customer demand. In some cases, that capability will be delivered via the Internet, as in many modern VPN installations. In other cases, it will not. It is crucial therefore to examine not only the Internet, but other access and transport mechanisms as well, including the burgeoning, high-bandwidth alternative local loop arena as well as *wide area network* (WAN) transport technologies such as frame relay, ATM, and native IP.

It is clear that many companies are intent on becoming the end-to-end solution for all the challenges that might confront a customer. Their acquisition models indicate a strong future focus on IP, but a current focus on ATM and circuit switching, as well as high-speed local access technologies such as DSL, cable modems, and high-speed, wireless local loops. More than anything else though, they are focused on that ephemeral yet critical thing called customer service. Without that, the best technology in the world becomes meaningless.

HOW TO USE THIS BOOK

This book examines all three components of the convergence phenomenon, but its primary goal is to help readers understand how the convergence of native telecommunications technologies and the companies that create them provide the fundamental underpinnings of customer service. Consequently, it is not a book targeted exclusively at the telecommunications industry, but is rather designed to help every industry make sense of convergence, and the fundamental role that telecommunications plays today.

It is divided into three primary sections: technology convergence, company convergence, and service convergence. In each section, we examine that particular facet of the convergence phenomenon by "dissecting" specific examples with an eye on the business implications of each. In the technology convergence section, we examine in detail the access and transport technologies that are germane to the ongoing evolution of electronic business. In the section on company convergence, we study the "business molecules" that have resulted from the calculated mergers and acquisitions that have so effectively characterized the telecommunications industry in the last three years. Finally, in the services convergence element, we pull it all together and consider how the combination of technological leadership and company ability results in the best possible customer service offering.

This book has a number of unique characteristics that distinguish it from others. First, although it is a technical book and does discuss telecommunications technology in reasonable detail, it does not attempt to teach the reader the first name of every bit in the bitstream. Myriad excellent books out there that accomplish that task superbly well. Instead, this book discusses the major technologies at a higher level but focuses on the business implications of the technologies. For example, if you are in the banking industry, what changes in computer technology do you need to be aware of that could impact the way you do business and assess customer service requirements? If you are in financial services, how do evolving local loop technologies affect you and your customers' abilities to do business with you? Will health care be affected by changes in WAN options that frame relay and ATM deployment offer? To what degree will the burgeoning growth of the Internet affect all of these industries? And how does the ongoing company convergence within the telecommunications marketplace affect the convergence activity in other industries such as automotive, health care, and financial services? These are the kinds of questions this book attempts to answer.

Second, the biggest challenge to writing about telecommunications technology lies in the very medium of the book you are reading. Telecommunications is the most dynamic, fast-changing industry in history, and any attempt to record the state of affairs of that industry on a non-changing medium like paper becomes something of an exercise in futility. We joke about the fact that because the technology changes so rapidly, such books are often out-of-date before they hit the bookstore shelves.

There is an element of truth to this, so to counter the fact, we have decided to publish this book and link it to a dedicated Web site. The material printed in the book has been carefully written to minimize the velocity at which its content will become dated. The Web site, located at www.shepardcomm.com/convergence, contains hundreds of references to the many aspects of the convergence process, carefully catalogued and easily searched. For example, if a reader wants to know more about a particular technology, he or she can simply go to the site and enter the term or terms they want to explore in the Web site's "Topic Search" field. A list of hyperlinked resources will be returned that can be perused at the reader's leisure. A summary of the Web site's contents is included in the Appendix of this book.

The site also offers a mechanism for readers to provide feedback to the author, to suggest the inclusion of additional content, and to engage in topical conversation with other readers who share common interests. The site will be maintained regularly to ensure that it is as current and topical as possible. It becomes, in effect, a knowledge portal. What better way to support a book built around the value of digital knowledge than to make that digital knowledge a part of the book?

I look forward to hearing from you. Enjoy the book.

Steven Shepard
Williston, Vermont
January 2000
Steve@shepardcomm.com

BEGINNINGS

In physics, there is a theory known as the "Big Bang" that attempts to explain the origins of the known universe. This recently postulated belief follows on the heels of work done by both Ptolemy in the second century, who argued that the Earth sits at the center of the universe while all other celestial bodies revolve around it, and that of Nicolaus Copernicus, who in the 16th century refuted the work of Ptolemy, claiming that the sun occupies the center of the universe and that the galactic bodies revolve around *it*. The Big Bang theory, suggested by Edwin Hubble early in the 20th century, contends that the universe was created following a cataclysmic explosion that occurred some 20 billion years ago, scattering cosmic debris in all directions that ultimately accreted into nebulae, planets, galaxies, and other cosmological bodies. He postulated that all matter was in a state of infinite density (the so-called "cosmic singularity") until the explosion occurred, resulting in the theory of a rapidly expanding universe and the basis of much of modern physics. This expanding universe theory also postulates that at some point, it will turn around and begin to collapse on itself.

Following the Telecommunications Big Bang of 1984 (usually called divestiture), that is precisely what has happened. When the telecommunications universe underwent its own cataclysmic explosion, during which AT&T was broken into a collection of companies, the many small corporations created in the aftermath remained independent for nearly 10 years but have now begun to accrete. The original 22 *Bell Operating Companies* (BOCs) were initially collapsed into seven *Regional*

Bell Operating Companies (RBOCs) and directed to play in the local services game and only in the local services game. AT&T was told to become a long-distance player, one of three created in the regulatory explosion. MCI and Sprint arose as powerful players in their own right, and a host of other corporate nebulae arose from the cosmic dust of divestiture to occupy places throughout the telecommunications universe (see Figure 0-1).

Now, however, it has become obvious that Hubble was right. The original seven RBOCs have collapsed to three (Bell Atlantic, which now owns NYNEX and GTE; SBC Corporation, which owns Pacific Bell, Nevada Bell, SNET, and Ameritech; and Bellsouth. USWest was acquired in 1999 by Qwest). The plethora of equipment manufacturers that grew in the wake of divestiture has entered into a feeding frenzy, acquiring each other at a frantic rate. In the last three years, Lucent, Nortel, and Cisco have acquired a diversity of companies as they struggle to reinvent themselves in the burgeoning marketplace.

Similarly, technologies and networks have begun to merge, while the services provided over those networks and by those companies have begun to blur. We have returned to the concept

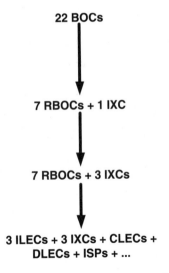

22 BOCs

7 RBOCs + 1 IXC

7 RBOCs + 3 IXCs

**3 ILECs + 3 IXCs + CLECs +
DLECs + ISPs + ...**

FIGURE 0-1   The evolution of telecommunications in the U.S.

of the network as a cloud. It doesn't matter what's in the cloud as long as it provides the services the customer requires. There is one difference, however. Before 1984, the customer had minimal choices. They took what the service provider offered because there was effectively only one provider. Today they have choices, and if the cloud fails to perform, the customer picks another cloud. Furthermore, the technology was the market driver in the years immediately following divestiture. The industry's research organizations (Bell Northern Research, Bell Laboratories, Xerox PARC, Bellcore, which is now Telcordia, and a host of others) rolled out innovative technology to the market, which would then scuttle away to build applications around those technologies. Today things are a bit different. The technology no longer leads the parade; the market does. The marketplace dictates applications, and for the first time, the technology follows the market as it races to accommodate application requirements.

There is an ancient Asian curse that wishes upon the accursed the desire that "they live in interesting times." By that measure, the telecommunications industry as a whole, and the customers it serves, have been roundly cursed. The changes that started in 1984 with the divestiture of the Bell System and the creation of the modern telecommunications industry continue apace, and the appearance is of an industry mired in chaos. And while there is a certain element of truth to that, it is also true that telecommunications is the fastest growing, most dynamic, and centrally critical provider of professional services in the world today. Try to identify an industry today that is not dependent upon telecommunications: I defy you.

THE STATE OF CHANGE

The changes that are underway are legion, and most are indicative of a healthy industry and a thriving marketplace. Generally, they fall into the same three convergence categories mentioned earlier: technologies, companies, and services. Consider the implications of the following changes for your own company:

Gozinta

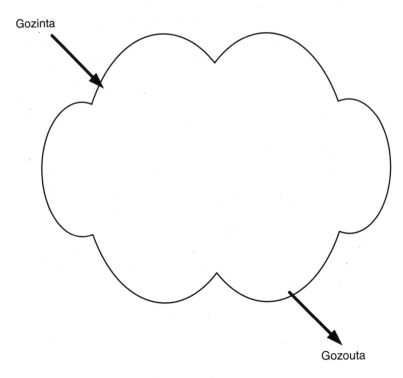

Gozouta

FIGURE 0-2   Generic network cloud

Customers no longer want to buy technology. There was a time not all that long ago when "service providers" sold raw bandwidth, ISDN, VPNs, ATM, frame relay, and other equally obtuse products. Today customers have evolved to the point that they no longer care about the underlying technology. In the same way that the telecommunications industry has become extremely competitive, so too have the vertical industries that the telecommunications providers serve. Those customers want to buy services that enhance their competitive positions and provide solutions to business problems. That the services are ultimately built upon a technological substrate is unimportant to them. What matters is that the services satisfy their ongoing business demands. The job of the service provider should be to screen the technology from the customer, providing "gozintas" and "gozoutas" to and from the network cloud that deliver connectivity and proper *quality-of-service* (QoS) levels as required (see Figure 0-2). This approach has several advantages. First, it gets the "service provider" out of the pure technology business, converting them instead into a service provider in the truest sense of the word. If technology is part of the service package they deliver, so be it. It is, however, merely a piece of the overall package.

Bandwidth has become a commodity. Commodities are typically defined as products that are differentiable on price alone. They are generally available from a wide variety of providers who sporadically share momentary market superiority when they find themselves offering the lowest price for the commodity at a given point in time. And although commodity markets tend to be hair-raising to operate in, they can also be extremely lucrative when played properly. After all, commodities are commodities because they are in universal demand.

In the last few years, a variety of companies have emerged on the bandwidth scene that have turned the market on its ear. Qwest and Level3, for example, have built DWDM-equipped optical networks with so much inherent bandwidth that they have redefined the pricing model, selling excess bandwidth at extremely low rates to the incumbent providers. By doing so, however, they threaten the livelihood of the legacy companies that have traditionally sold bandwidth as their primary, most lucrative product. These traditional players, the ILECs (formerly the RBOCs) and the IXCs, have historically provided access and transport as their principal product offerings, which fundamentally rely upon bandwidth as their raw material. As bandwidth becomes universally available and approaches commodity status, the protective fiefdoms of the ILECs and IXCs come under attack by the bandwidth berserkers who have the capacity and low per-unit prices to attack them where it hurts most. Furthermore, these companies often pose a dual threat to the incumbents by forming alliances with CLECs, who together have the ability to bypass the ILECs entirely. This is not new to the incumbents. They realize that if they are to remain competitive, they must have access to long-distance service that will enable them to offer end-to-end services in a distance-insensitive fashion.

Customers are demanding more bandwidth. Driven by inexpensive bandwidth, low-cost storage media, the rise of *Applications Service Providers* (ASPs) and appliance-based, feature-rich applications, customers are demanding significantly more bandwidth than ever before. Furthermore, the profile of the typical user is changing as telecommuters and *Small-Office-Home-Office* (SOHO) workers increase. As bandwidth-hungry applications become more common, the demand for broadband access will grow apace, requiring access providers to re-examine their infrastructure strategies. Furthermore, they will have to accept the fact that how they sell will be at least as important as what they sell. Broadband will be differentiated on customer service and the degree to which its marketing is successful.

The role of the service provider has changed dramatically. The term "service provider" has always been synonymous with the providers of access and transport—no longer. Today the term means exactly what the words imply: that these companies must evolve to become providers of a full suite of services as defined by the business needs of their customers. If they fail to do so, they risk loss of marketshare.

These companies may find themselves paying a great deal of attention to the content arena. Although there is certainly money to be made in the provision of access and transport—after all, connectivity is central to the success of any information-dependent industry—the ultimate goal of players in the telecommunications game should be to become a full-service provider, offering a spectrum of services that combine universal access, transport, a wide range of service quality levels, and diverse content. One problem with content, of course, is the volume of it that is available. Recent studies have shown that even the most comprehensive search engines only scrutinize a tiny percentage of the total number of Web pages that are online in spite of the fact that they often return lists of "found" sites following a query that number in the hundreds of thousands. A valuable service, then, would be one that provides some form of intelligent filtration of the information based upon a predefined subset of user preferences. Industry and activity-specific portals have recently emerged that provide this service by identifying and packaging content in such a way that it satisfies the requirements of specific industry segments. A winning combination, then, might be a portal service combined with an access and transport provider.

Managerial structures are changing. Modern corporations are undergoing their own big bang as they evolve from centrally managed, hierarchically controlled, top-down entities into a loose collection of self-managed organizations that operate efficiently in a distributed, peer-to-peer fashion. As the seat of operational power migrates from the pinnacle of the corporate pyramid to the lowest levels of the managerial ranks, decision-making is moved to lower levels as well. As a consequence, operational decisions are made faster and more accurately by those who must ultimately carry them out.

The networks that support the evolving corporations are evolving. The original corporate networks would have made Ptolemy proud. At the center of the computing universe was a large, expensive mainframe computer that consolidated all digital decision-making power in one place. Hierarchically below the mainframe was a front-end processor that managed communications between lower-level devices and the mainframe. Below the front end were cluster controllers and the so-called "dumb terminals," access devices that had all the intelligence of a television set with an attached keyboard. In this model, all roads led to the mainframe, the seat of decision-making power in the network.

IBM's original *Systems Network Architecture* (SNA) typified this model. It was extremely secure, supported a large population of users and enormous databases, and was the perfect technological complement to the centralized managerial style of most major corporations in the early 1970s when it became available.

Today mainframes are still very much in evidence, but are no longer necessarily at the center of the computing universe. Instead they have become one more computing resource attached to a corporate network that is no longer hierarchical. The network is now flat, and computing power has been distributed among a collection of extremely powerful, far less expensive devices.

It is interesting to note that the network has evolved in lockstep with the corporation. As the corporation's hierarchical structure has been eroded by the weathering effects of marketplace evolution, so too has the network that supports that evolution. As we might expect, SNA now supports peer-to-peer network architectures.[1]

The local loop is evolving. For years, the traditional twisted pair local loop provided adequate connectivity services for most applications. Today, however, as demands for bandwidth are at an all-time high, other, more innovative solutions have emerged, including 56K modems, ISDN, xDSL, cable modems, *Hybrid Fiber Coax* (HFC), *Switched Digital Video* (SDV), wireless options, and a variety of others.

Wireless access technologies are becoming centrally important. Not only do they provide the freedom of mobility that business and society demand, in many cases, wireless technologies are less expensive to deploy than traditional wired infrastructure solutions. As the technologies evolve in terms of both bandwidth capability and payload quality assurance, and as the regulatory strictures that govern wireless deployment become more liberal, technologies such as LMDS, MMDS, *Global System for Mobile Communications* (GSM), and *third generation* (3G) cellular will become mainstream options for access. According to the *International Telecommunication Union* (ITU), the number of wireless subscribers will surpass the number of fixed, "wireline" subscribers by 2008.[2] Furthermore, the complement of wireless applications will include much more than simple voice. To it will be added Web

---

[1]It should be noted that SNA is still in widespread use, but has evolved to accommodate the less-hierarchical requirements of modern networks. Such advancements as *Advanced Peer-to-Peer Networking* (APPN) and *Advanced Program-to-Program Communication* (APPC) are examples of this evolution.

[2]Source: ITU Telecommunications Indicators Database.

access, corporate *local area network* (LAN) and database access, fax, paging, e-mail, and others. And with the growing popularity of mobile access devices such as the Palm Pilot, it is only a matter of time before cell phones become an integral part of the device. The Palm VII already offers wireless access to e-mail and the Web. Telephony represents a simple add-on.

The fabric of the transport network is changing. The switched telephone network is designed to transport analog voice and lower-speed data. Implemented over an infrastructure of circuit switches such as the Lucent Technologies 5ESS, the Nortel Networks DMS-100, the Siemens EWSD, or the Ericsson AXE, the network provides dedicated pathways for each call. As the nature of the transported information has changed to include more bandwidth-intensive data, and as the length of the average call has increased from approximately 3 minutes to 10 minutes or more, the network infrastructure has proven inadequate. Consequently, new switching technologies have been introduced that are packet-based and more capable of satisfying the requirements of a data-centric network. The Lucent Technologies 7R/E, the Cisco Gigaswitch, and the Nortel Networks Succession are examples of these new platforms. Designed in recognition of the realities of the current marketplace, they can be installed as additions to the existing switch infrastructure to add capability incrementally, as the market demands it. The existing circuit switch simply becomes part of the evolved modular switch, which now offers both voice and packet-switched data.

IP is ascending to the throne of the protocol monarchy. The *Internet protocol* (IP) is the most widely deployed protocol on Earth, with the possible exception of the SS7 suite that controls call setup on the world-wide circuit switched telephone network. IP's ubiquity, along with its universally accepted addressing scheme and the fact that it is a packet-based architecture, has enabled it to insinuate itself into LANs and WANs, network operating systems, and strategic deployment plans for the evolving telephone network. It will undoubtedly become the primary network layer protocol for global, full-service networks, but its time has not yet come. It must continue to evolve technologically if it is to provide QoS discrimination for mixed traffic networks.

The Internet will continue to be a focal point for the evolving network-centric world. Although the Internet is not yet the "network of networks" as it has been predicted to become, it is serving as the paradigmatic indicator of what is to come. Because it is accessible to everyone and can therefore serve as a low-cost application and capability proving ground, it has served to put networking into the forefront of the business world. A day rarely passes when *The Wall Street Journal* does not contain multiple articles that revolve around networking capability.

Furthermore, the Internet is redefining the business world as electronic commerce and electronic business become reliable, trusted, and lucrative channels for conducting commercial activities on a global basis. By eliminating the barriers created by distance and size, the Internet reinvents the rules of the competitive game, enabling small, far-flung companies to compete on equal footing with commercial behemoths. Competitive advantage becomes more a factor of the ability to read the tea leaves of the marketplace and provide customer service than physical size and presence.

Management roles are evolving, sometimes in unusual ways. During the late 1960s and early 1970s when corporations first started to acquire mainframe computers and build networks around them, *information technology* (IT) resources were typically placed under the control of the *Chief Financial Officer* (CFO). This made sense because the purchase, after all, represented a $30 to 40 million investment, and who better to manage such a large capital budget than the finance organization? Furthermore, the initial applications that ran on those early systems supported financial processes—payroll and accounting.

As time went on, it became clear during the late 1970s that the applications were expanding beyond the realm of the financial organization. Because of this cross-domain evolution, a new officer position was created to administer the IT resources. The newly-appointed *Chief Information Officer* (CIO) became responsible for the strategic deployment of information resources throughout the corporation that served to facilitate the company's business activities.

In the mid-1980s, a number of natural and human-caused disasters shifted management's focus away from computing resources toward a focus on network resources. In the aftermath of central office fires, flooded cable vaults, earthquakes, and repeated instances of "backhoe fade," it became clear that a corporation can have the most advanced and widely deployed computer resources possible, but without an adequate, survivable network in place to ensure that employees and customers have the ability to reach the information contained in those computer systems, they quickly become useless. Consequently, the *Chief Networking Officer* (CNO) began to appear in many corporations, tasked with the responsibility to ensure that corporate networks were deployed in a redundant fashion and that no single component or full-span failure would result in the inability of users to reach systems.

For the next few years, corporations flirted with various managerial configurations, including such titles as *Chief Technology Officers* (CTO) and a variety of others. All of them, however, remained focused on the preservation of data resources.

This perspective remained static until the 1990s when a new officer began to appear on corporate battlefields. This position, the *Chief Knowledge Officer* (CKO), has a different set of responsibilities than his or her predecessors and peers. Instead of a focus on computer resources, network deployment, or database integrity, this officer's responsibility is to ensure that knowledge is properly disseminated throughout the corporation. This represents a major departure from prior managerial philosophies. Knowledge is a resource that is difficult to quantify, identify, or manage. Yet the highest corporate managerial ranks recognize intuitively that it is critical. Knowledge is not data; it is not information. Rather, it represents an understanding or awareness of data and information that comes from experience and familiarity with the environment in which one operates. Knowledge of customer behaviors, then, can be extremely valuable to a corporation under competitive siege and can make the difference in a marketplace such as telecommunications where many of the products and services are becoming commodities. In some corporations today, there is now a Chief Change Officer, responsible for monitoring E-business and the effect it and the information it depends on have on the business.

The knowledge-based corporation is a reality, and the knowledge worker is on the rise. Corporations now realize that the more they know about their customers and their customers' habits in the marketplace, the more competitive they can be when vying for those customers' attentions. Furthermore, workers whose principal contribution to the corporation is knowledge are proving to be enormously valuable. Related to this, corporations are reawakening to the importance of *Operations Support Systems* (OSSs) that help them manage internal processes related to security, system and network configuration and performance, fault detection and resolution, and resource management.

In support of this realization, a new breed of support organization has emerged on the scene, offering a collection of innovative services to the competitive corporation. These are known as *Enterprise Resource Planning* (ERP) services, and their goal is to help companies assess customer requirements and behaviors by teaching them how to analyze the contents of their own corporate databases. This activity, often referred to as *data mining*, is proving to be a strategic weapon in the competitive arsenal of most corporations.

Equally important is the activity known as *Customer Relationship Management* (CRM). In CRM, companies build upon the results of ERP activities to ensure that their own business activities dovetail properly with those of the customer. Companies such as SAP, Sterling, Peoplesoft, Brio, Hewlett-Packard and others have become well-known providers of ERP and CRM services.

Intelligence is migrating from the center of the network out toward the user's access device. As computing power becomes less and less expensive, it will be used more universally to distribute network intelligence away from a small number of centralized, shared nodes toward a larger number of distributed devices, often referred to as user appliances. According to the Semiconductor Industry Association's *World Semiconductor Forecast 1998–2001*, the total number of semiconductor chips installed in devices other than computers will grow precipitously.

The players in the game are changing. Once the exclusive domain of the traditional local and long-distance players, telecommunications has evolved into a who's who of access and transport providers. The ILECs and IXCs are still the leading providers, but nipping at their heels is a pack of customer-hungry alternatives who represent a growing, serious threat of competitive turbulence to the calm waters of the incumbent marketplace. These players include cable television companies whose recently upgraded networks now enable a high-speed, digital, two-way transport of information. This includes telephony and Internet access; satellite providers, who, thanks to regulatory changes, can now offer two-way Internet access in addition to their traditional programming lineup; CLECs, who are building their own networks to compete head-to-head with incumbent providers; ISPs; fixed wireless providers who sell alternative local loop solutions; and a host of others. Competition has run amok, and although the resulting market is chaotic for the companies who play there, customers benefit from the competitive pricing and service diversity that results.

The traditional, LATA-bound market is a thing of the past. Today corporations must think well beyond their traditional geographic boundaries to create strategies that will enable them to enter the global markets made possible by the Internet and other global networks. In September 1995, *The Economist* published an article entitled "The Death of Distance" in which the author observed that because of the global connectivity made possible by the Internet and the World Wide Web, the physical location of a business is no longer important. In his book *Being Digital*, Nicholas Negroponte discusses the fact that we are entering a world in which businesses make money by moving bits around rather than by moving atoms, or information instead of physical goods. Both of these observations point to the fact that the marketplace has become the entire world.

SUMMARY

So why do we care about convergence and the evolving world of telecommunications? We care because this collapse, this convergence of all things technological, is an indication of what's important in most industries. Customers don't care as much about underlying technologies anymore. They care about the services that the technologies provide, because those services give them the differentiation they need to be competitive. The convergence phenomenon is going on in other industries as well. In banking, health care, and energy, large corporations are merging to become larger. The phenomenon is as much managerial and service-oriented as it is technological in nature and illustrates one critical point: telecommunications technologies are central to the success of all these mergers and acquisitions. The separation between the merging companies and the digital glue that binds them is so minimal that "the company" and "the network" have become largely indistinguishable.

No doubt about it. Telecom has become mainstream, the competitive linchpin upon which most corporations hinge.

RESOURCES

BOOKS

Davis, Stanley M., Christopher Meyer and Stan Davis. *Blur: The Speed of Change in the Connected Economy.* Addison-Wesley; Reading, MA, 1998.

Tapscott, Donald, Alex Lowy and David Ticoll. *Blueprint for the Digital Economy.* McGraw-Hill; New York, 1998.

Tapscott, Donald. *Growing Up Digital: The Rise of the Net Generation.* McGraw-Hill; New York, 1998.

Helegesen, Sally. *The Web of Inclusion: Building an Organization for Everyone.* Doubleday, 1995.

Evans, Philip and Thomas S. Wurster. *Blown to Bits: How the New Economics of Information Transforms Society.* Harvard Business School Press; Boston, 2000.

Peters, Tom. *The Circle of Innovation: You Can't Shrink Your Way to Greatness.* Vintage; New York, 1999.

Porter, Michael. *Competitive Advantage: Creating and Sustaining Superior Performance.* Free Press; Boston, 1985.

Negroponte, Nicholas. Being Digital. Alfred A. Knopf; New York, 1995.

———. *The Connected Society: Winning the New Battle for the Customer.* Ernst & Young; 1999.

WEB RESOURCES

The New Information Industry Tutorial. Web ProForums; International Engineering Consortium. No date provided. `www.webproforum.com/new_info/index.html`.

The Rise of the Stupid Network. Isenberg, David; 1997. `www.camworld.com/stupid.html`.

Cisco Releases Study Measuring Jobs and Revenues Tied to the Internet Economy. June 10, 1999. Company press release.

The Internet Economy Indicators. June 11, 1999. `www.internetindicators.com/features.html`.

ARTICLES

3Com Corporation. "In Phase With the Future." *Net Age*, Q3 1999, page 6.

Breidenbach, Susan. "Got the Urge to Converge?" *Network World*, September 27, 1999, page 51.

Briere, Daniel and Christine Heckart. "What Organized Crime and Convergence Have in Common." *Network World*, August 1999.

Cross, Kim. "B-to-B, By the Numbers." *Business 2.0*, September 1999, page 109.

DeVeaux, Paul. "Fighting Among the Switch Vendors." *America's Network*, August 1, 1999, page 26.

Drucker, Peter F. "Beyond the Information Revolution." *The Atlantic Monthly*, October 1999, page 47.

Figueredo, Ken and Brian Toll. "Are You Ready for Convergence?" *Network World*, September 13, 1999, page 99.

Fox, Loren. "Another Face for Venture Capitalism?" *Upside*, October 1999, page 127.

Gross, Neil. "Ideas for the 21st Century." *Business Week*, August 30, 1999, page 132.

Isenberg, David. "Mother of all Disruptions." *America's Network*, July 15, 1999, page 12.

Liebmann, Lenny. "Bandwidth: Raw Material for the new Economy." *Communications News*, April 1999, page 84.

Lieberman, David. "Bridging the Digital Divide." *USA Today*, October 11, 1999, page 3B.

Rybczynski, Tony. "Follow the Leaders." *Communications News*, April 1999, page 26.

Sawhney, Mohanbir and Steven Kaplan. "Let's Get Vertical." *Business 2.0*, September 1999, page 85.

Schlender, Brent. "The Real Road Ahead." *Fortune*, October 25, 1999, page 138.

Steinert-Threlkeld, Tom. "The Net: Not Just Data Anymore." *Internet 2002*, August 31, 1998.

Tolly, Kevin. "Converged Telephony in 1999?" *Network World*, May 31, 1999, page 24.

THE TECHNOLOGIES

As the foundation layer of the convergence troika, telecommunications technologies can be subdivided into two interrelated segments: *access*, which includes not only the actual circuit used to connect user devices to the network, but the access device technologies themselves, and *transport*, which describes the fabric of the *wide area network* (WAN). Two other technology areas, *signaling*, the protocols and procedures used to establish, maintain, and terminate voice, data, and video transmissions across the network, and *network management*, the set of procedures used to operate, maintain and provision network resources, will be discussed later.

We will approach our examination of communications technology from the point of view of each of these because in reality, all have some bearing on the capability of the network to provide services to the customer. Access, including the device(s) used by the customer to gain access to the network itself, addresses local loop technology options that a customer might employ to enter the network cloud, such as traditional twisted pair, ISDN, *x-Type Digital Subscriber Line* (xDSL), cable modems, and wireless services. This region of the network is particularly important because it touches the customer. Whoever controls access may well control the customer, as evidenced by the fierce battle that is underway between the telephone companies and their emerging arch-competitors, the cable providers.

Transport defines the set of technologies that make up the fabric of the network cloud, including circuit switching, traditional packet, frame relay, *Asynchronous Transfer Mode* (ATM), and IP. As the marketplace evolves to demand more in the way of wide-area, end-to-end services (particularly data), the nature of the transport fabric becomes rather critical, particularly in converged networks where it must capably transport a diversity of traffic types, all with appropriate levels of service quality. Furthermore, it is becoming evident that a pure technology network is probably not in the cards. Most likely, the network of the foreseeable future will comprise a hybrid of multiple technologies that together are capable of supporting the diverse transport requirements of an increasingly complex set of customer applications. Ideally, of course, the customer will remain blissfully unaware of this technological cornucopia.

Signaling is one of the most important yet least understood aspects of telecommunications networking. In order to establish, maintain, and tear down a connection, some form of signaling protocol must be employed. Not only must it accomplish the three functions just mentioned, it must also be able to dip into distributed databases to retrieve subscriber information and establish the connection according to the information it finds there. For example, *Signaling System Seven* (SS7) employs customer databases known as *Service Control Points* (SCPs) that store calling feature information, credit card data, 800/888-number translations, and a plethora of other data points that can be used by the network to quickly and accurately establish a connection for the subscriber while invoking a preferred set of features. Catalog companies routinely rely on this technology to give themselves the ability to greet a caller by name and discuss that caller's recent purchase history. It gives fast-food delivery companies the ability to track customer purchase histories and target those who ordered once and never called again. They can identify those callers and send them a special promotion to try to win them back. Signaling is the technology that makes this possible.

In the multimedia environment, signaling becomes centrally important because it enables a multisession call to be established. For example, a voice conversation between two people may reach a point where the callers need to invoke a videoconference session. Through multimedia signaling, they can easily do so without affecting the ongoing voice call. Furthermore, they can add other sessions such as fax or data as required simply by requesting that the signaling network identify the necessary bandwidth, reserve it, qualify its capability to provide acceptable *quality of service* (QoS) and establish the path between the identified endpoints. In the evolving call center environment, this ability to establish multimedia sessions is an important part of the customer service equation.

Network management, often referred to in telephone company parlance as *Operations, Administration, Maintenance, and Provisioning* (OAM&P), provides the toolset required for network hardware and software to deliver what the customer demands in the way of network services. *Operations* is an umbrella term that includes the other three. *Administration* denotes internal processes required to keep the network functioning properly. *Maintenance* is the task of monitoring the network, detecting faults when they occur, and repairing them. *Provisioning* is the process of designing and building network infrastructure throughout the operations area so that when customers request service, the network is in place to deliver it in a timely manner. All four of these functions will be examined in detail.

ACCESS TECHNOLOGIES

Historically, "access" has referred to that subset of the network that enables the customer to connect to the wide area transport fabric, whether it be the *Public Switched Telephone Network* (PSTN), an ATM, a frame relay network, the Internet, or some other solution. In its earliest incarnation, access referred to the local loop, the twisted copper pair that has historically provided connectivity between the customer's access device and the network cloud.

When the telephone network was originally conceived in the late 19th century, it was designed to transport the limited bandwidth requirements of the only traffic type in existence at the time, voice.[1] By the middle of the 1920s, thanks to the work done by such Bell Labs luminaries as Harry Nyquist, it was well known that the human voice comprised a rich spectrum of frequencies that ranged from 20 Hz to approximately 20 KHz. It was also known that although this broad spectral range contributed to the extreme richness and timbre of the human voice, it was not required to recognize or understand the person on the other end of the circuit. This was actually a good thing. If voice required the provisioning of 20 KHz of bandwidth to every telephone call, the cost to build a network capable of doing that would be enormous and highly impractical. As a consequence, network engineers determined that as long as each telephony conversation was given four KHz, a far cry from 20,000, the person at the other end of the line could be both recognized and understood. For the delivery of voice, this proved to be adequate.

Time went on, however, and the legacy analog network evolved. By the late 1950s, work was being done on digital transmission, and in 1962 the first T-Carrier system was installed to carry digitized voice. This was a great step forward because it allowed network engineers to take advantage of the magic of digital transmission. Specifically, it meant that transmission lines could be made "cleaner," and that pair gain was achievable, that is, it was possible to transmit multiple conversations across a single facility.

Let us dispel a common misconception here. It is often said that digital facilities experience fewer errors than their analog counterparts. In fact, this is not true. Both forms of transmission suffer from the same impairments—impulse noise, crosstalk, thermal noise, harmonic distortion, and a variety of others. The difference lies in how they deal with those impairments. In order to understand that difference, we need to first conduct a short

[1]The story is well known. According to the misty memory of history, Alexander Graham Bell reportedly spilled a beaker of acid into his lap, causing him to say, "Come here, Watson, I need you." I submit to you that that is probably not what he said, at least not first, and Watson probably didn't need a telephone to hear him.

and painless discussion of the difference between signaling and content, that is, the difference between the way in which we represent the information on the transmission line and the information itself.

In common network parlance, the word "analog" means "continuous." An analog wave is a wave that is continuously varying in time. Imagine if you were to use an analog light meter, one with a needle, to create a graph of the intensity of sunlight from sunrise to sunset. The graph would be continuously variable over the 12-hour period and would look like the drawing in Figure 1-1.

Now do the same thing, but this time use a digital light meter, the kind with digital readouts. As you graph the intensity curve, it will not be continuously variable over time, but will in fact comprise a series of discrete "stairsteps," as shown in Figure 1-2, that represent the measured intensity of the sunlight.

The analog wave represents an infinite series of data points to convey the required information, while the digital (the word, by the way, means *discrete*) "wave" comprises a far more limited number of data points to convey the same information. Obviously,

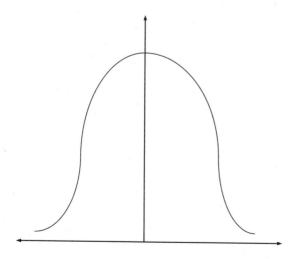

FIGURE 1-1   The analog intensity of sunlight

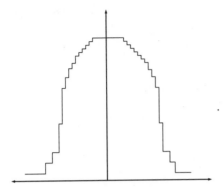

FIGURE 1-2   The digital intensity of sunlight

the digital wave represents an approximation of the actual intensity of the sunlight at any point in time, but it should be obvious to the reader that if enough data points are captured, the digital wave will do an adequate job of recreating the information.

The techniques described here represent *signaling techniques*. Analog signaling uses a continuously varying waveform to represent the information transported, while digital transmission uses a discrete collection of data points to do the same thing. Both are effective, and although it could be argued that the digital technique is really only an approximation of the actual content, we know from further work done by Nyquist in the late 1920s that there is an optimal number of data points that must be generated to create a faithful representation of the original signal.[2]

Analog waves can be modified in a variety of ways to cause them to carry information. Specifically, three characteristics of the wave can be modified: the amplitude, the frequency, and the phase. The amplitude is a measure of the loudness of the signal; it can be high or low. The frequency, similarly, can be

---

[2]Specifically, Nyquist postulated that as long as the original signal is sampled at a rate that is twice the highest frequency in the original signal, the resulting series of discrete samples will be adequate to reproduce the original signal. For example, the upper end of the so-called voice band is 4 KHz; according to Nyquist, then, sampling 8,000 times per second will result in a completely accurate and faithful representation of the original signal. This sampling rate calculation is known as "Nyquist's Theorem."

high or low. And the phase, a somewhat more complex concept, can be modified to represent the "location" of the wave form relative to a reference point at any moment in time.

It is important for the purposes of this discussion to remember that we are still talking about how we represent information in a transmission facility, not the nature of the information itself. In the original analog telephone network, analog waves (the actual voice impinging on the microphone of the telephone) caused a current to vary in analog fashion, which in turn passed through the fabric of the network to the telephone on the other end. At the receiving phone, the same varying current caused the speaker in the handset to vibrate analogously, creating sound that the receiver was able to hear. The point is that the information going into the network was analog in nature, it was transmitted in analog format, and it was delivered as an analog wave.

Now let's replace the analog network with a digital network. In this case, the incoming signal (voice) is analog; it is converted to a series of discrete samples that represent the original analog signal; the samples are converted to digital approximations, the bits are sent into the netowrk, and at the other end the bits are converted to analog samples from which we reconstruct the original signal for delivery to the receiving customer's ear. Please understand: the information being transported is analog. By choosing to digitize it for transmission, however, we represent it in a digital fashion. The information itself is still analog, and always will be.

So why did we choose to go through this technological nosebleed? Because digital and analog technologies currently coexist peacefully and will for some time to come. Just because information is transported over a digital network does not mean that the information itself is digital. Consider, for example, the relatively new technology known as *Asymmetric Digital Subscriber Line* (ADSL), which enables massive amounts of bandwidth to be delivered across a two-wire local loop. In ADSL, in spite of the fact that the information being transported is often digital (hence the name), the actual transmission facility is analog. The devices at either end of the circuit are in fact modems, not digital transmission devices.

So back to our original point about the relative "goodness" of analog and digital facilities. Digital facilities are not "cleaner" than their analog predecessors. In that regard, there is no difference. The difference, and the relative goodness of the two, lies in how they deal with the impairments that affect the data they transport. In both systems, signals have a tendency to degrade or weaken over distance. They also have a tendency to be affected by spurious noise signals, and those noise signals are unfortunately cumulative.

In analog systems, where there is a continuous, infinite stream of useful data points, errors can often be indistinguishable from the data itself. A 30-ms impulse noise hit, for example, can look like a data point and therefore result in an error in the information stream. Furthermore, as the actual signal weakens over its transmitted distance, the level of the desired signal weakens relative to the level of the accumulated noise, the *signal-to-noise ratio*. At some point in the facility, the signal weakens to the point that it must be amplified. The problem with amplification in analog systems is that the amplification process is non-discriminatory. The amplifier amplifies whatever it is fed, which includes both the original signal and the noise it has accumulated along the way. The result of this is a louder, noisy signal, clearly not the most desirable outcome.

In digital systems, things happen a bit differently. Because digital transmission systems rely by their very nature on a limited set of discrete data points (zeroes and ones, or some combination thereof), it is easier for digital amplification equipment (usually called *repeaters* or *regenerators*) to identify the original signal and filter everything else out, resulting in a recreation of the original signal (hence the term regenerator). Therefore, we see that digital networks are not cleaner than their analog counterparts; they're just better at dealing with the dirt.

Of course, analog systems can provide reliable transport services in myriad ways. Coaxial cable systems are often analog today, but because the physical facility is shielded so well, errors rarely occur. By the same token, complex error correction and detection protocols have been developed that do a good job of eliminating the problems of noise in analog transmission systems.

So now we understand the difference between signaling and content. Signaling describes the manner in which we represent the content that is being transported across the network, while content describes the information itself, which can be analog (such as voice) or digital (such as the stream of zeroes and ones that spew from the serial port of a PC).

TRADITIONAL ACCESS TECHNOLOGIES

Because the human voice only requires four KHz of bandwidth to achieve reasonable transmission quality, the analog telephone network, including the local loop, has been designed to allocate exactly that much bandwidth to each voice channel. In reality, that 4KHz allocation includes some guard band on either side of the voice channel to prevent crosstalk. 3.1KHz are actually given to each channel. Since the late 1960s, the inner core of the network has been progressively digitized starting with the introduction of T-Carrier in 1962 for central-office-to-central-office transports and the arrival of the first digital switches in the 1970s. The local loop, however, has largely remained analog.

In 1981, IBM turned the world on its ear with the introduction of the PC, and in 1984, the Macintosh arrived, bringing computing power to the proverbial masses. Shortly thereafter, hobbyists and hackers[3] began to take advantage of emergent modem technology and create online databases, the first bulletin board systems that enabled people to send simple text messages to each other. This accelerated the modem market dramatically, and before long, data became a common component of local loop traffic. Similarly, the business world found more and more applications for data, and the need to move that data from place to place also became a major contributor to the growth in data traffic on the world's telephone networks.

At this point, data did not represent a problem for the bandwidth-limited local loop. The digital information created

---

[3]It should be observed that the original hackers were not viewed as evil creatures, but rather as "tinkerers" who pushed the creativity envelope.

by a computer and intended for transmission through the telephone network was received by a modem, converted into a modulated analog waveform that fell within the four-KHz voice band, and fed to the network without incident. The modem's job was (and is) quite simple. When a computer is doing the talking, the modem must make the network think it is talking to a telephone.

Over time, modem technology advanced, enabling the local loop to provide higher and higher bandwidth. This increasing bandwidth was made possible through clever signaling schemes that enabled a single signaling event to transport more than just a single bit. Consider the following example. If we use amplitude modulation to represent digital data, we might have a system that uses a high-amplitude (loud) wave to represent a one and a low-amplitude wave (soft) to represent a zero. In this system, each signal represents a single bit, which means that if we are limited to the four-KHz voice band, our bit rate is similarly limited. In fact, Nyquist's theorem also demonstrated that the fastest rate that we can signal at in a bandwidth-limited system is at a rate that is no faster than twice the bandwidth of the channel in which we are operating. In the voice band, then, the fastest signaling rate we can ever achieve is 2×3.1 KHz, or 6,000 signals per second.

So why do we care about this esoterica? We care because the signaling rate has a direct impact on the achievable bit rate, and that is a central service component. If the width of the channel limits the signaling rate, then it obviously also limits the bits represented by those signals. So what are the available options? If the channel bandwidth limits the signaling rate, then obviously a bigger pipe means more signals. The problem is that bigger pipes mean more money, because bandwidth is expensive. So, although this may work, it is not necessarily the most desirable option. Besides, the voice band is limited to four KHz, and the switches are looking for information to come out of that limited channel. So, a bigger voice band is not a viable option.

Alternatively, we might create a way to represent multiple bits on each transmitted signal. In fact, this is precisely what modern modems do. Consider our previous example in which we described a system where a single bit was represented by each discrete signal—high amplitude connoted a one, low amplitude a zero. What happens if we now add a little complexity? Instead of relying purely on amplitude modulation, we now add frequency modulation as well, as shown in Figure 1-3. As a consequence, we now have four definable states that we can use to represent information instead of two: high amplitude, high frequency; high amplitude, low frequency; low amplitude, high frequency; and

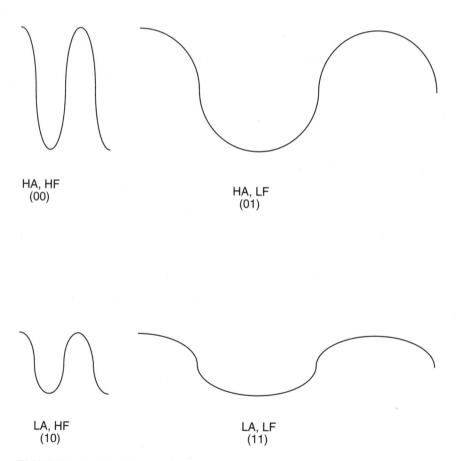

HA, HF
(00)

HA, LF
(01)

LA, HF
(10)

LA, LF
(11)

FIGURE 1-3   Waveform modulation

low amplitude, low frequency. Instead of carrying a single bit, each of these can now carry two bits, which means that our bit rate is now twice our signaling rate, often referred to as *baud*.

Consider the consequences of the following. If the signaling rate is limited by the bandwidth of the channel in which we are required to operate, we can overcome this limitation by convincing each signal to carry multiple bits, thus achieving higher bit rates in spite of a bandwidth-limited channel. Many examples of this exist, ranging from the inner workings of 9,600-bit-per-second modems that encode 4 bits per signal to cable systems capable of encoding as many as 10 bits per signal. Needless to say, at these encoding levels, the degree to which noise is present can have a deleterious effect on cleanly transporting data, so noise reduction techniques are of paramount importance.[4] Some modems employ highly sophisticated error detection and correction techniques, such as *trellis coded modulation*. Other systems rely on the physical nature of the medium that they transport their cargo over to reduce the error rate. Coaxial cable, with its inherent shielding, is one example of this, and fiber, with its immunity to electrical interference, is another.

The analog local loop is employed for a variety of voice and data applications today in both the business and residence marketplaces. The new lease on life it enjoys, thanks to advanced modem technology as well as a focus by installation personnel on the need to build clean, reliable outside plant, has resulted in the development of faster access technologies designed to operate across the analog local loop.

This does not mean to imply that traditional, pre-existing technologies will disappear any time soon. Hundreds of thousands, if not millions, of modems are still installed in computers today that operate at speeds in the range of 33.6 Kbps. New machines, however, are routinely equipped with 56-Kbps

---

[4]Work performed by Claude Shannon, often referred to as *Shannon's Law*, concludes that the number of bits that can be encoded on a single signal is limited by the signal-to-noise ratio. In effect, the more noise, the less bits per signal.

modems and in some cases with ADSL and even cable modems. A discussion of these various access devices and the relative advantages and disadvantages they offer follows.

MARKETPLACE REALITIES

According to any number of both private and government-conducted studies, approximately 40 million households today host home office workers, including both telecommuters and those who are self-employed and work out of their homes. These workers require the ability to connect to remote *local area networks* (LANs) and corporate databases, retrieve e-mail, access the Web, and in some cases conduct videoconferences with colleagues and customers. The traditional local loop, with its bandwidth-limited capabilities, is not capable of easily satisfying these requirements with standard modem connectivity. Dedicated private-line service, which would certainly solve the problem, is out of the financial reach of many of these businesses or is not allowed by parent companies for remote workers. Other solutions are required, and these have emerged in the form of a new flock of access technologies that take advantage of either a conversion to end-to-end digital connectivity (ISDN) or expanded capabilities of the traditional analog local loop (xDSL or 56K modems). In some cases, a whole new architectural approach is causing excitement in the industry, such as *Wireless Local Loop* (WLL).

56-KBPS MODEMS

One of the most important words in telecommunications is "virtual." It is used in a variety of ways, but in reality only has one meaning. If you see the word virtual associated with a technology or product, you should immediately say to yourself, "It's a lie."

56-Kbps modems are good examples. They have garnered a significant amount of interest since they were introduced a few years ago. Under certain circumstances, they do offer higher access speeds designed to satisfy the increasing demands of bandwidth-hungry applications and increasingly graphics-oriented Web pages. The problem with these so-called 56K modems is that they are virtual modems in that they do not really provide true 56K access, even under the best of circumstances.

These devices are designed to provide *asymmetric bandwidth* with 56 Kbps delivered downstream toward the customer and significantly less bandwidth (33.6 Kbps) in the upstream direction. Although this may seem odd, it makes sense given the requirements of most applications today that require modem access. A Web session, for example, requires very little bandwidth in the upstream direction to request that a page be downloaded. The page itself, however, may require significantly more since it may be replete with text, graphics, Java applets, and even small video clips. Since the majority of modem access today is for Internet surfing, this asymmetric technique is adequate for most users today. In fact, tests show that users get approximately 20 percent greater throughput with a 56K modem than they get with 33.6 Kbps.

56K modem technology experienced something of a market rebellion when it was first introduced. Two industry players simultaneously introduced 56K standards that worked well but were incompatible with one another. Motorola, Rockwell, and Lucent collectively introduced their K56 Flex technology, while U.S. Robotics (now 3Com) introduced a standard called X2. The problem with this simultaneous, incompatible product introduction was that it occurred at a time when online usage was at an all-time high and the marketplace, particularly the ISPs, were looking to buy enormous volumes of modems to augment their modem pools and slake their customers' hunger for greater bandwidth. Unfortunately, because there were two mutually incompatible standards, both introduced by reputable firms, the modem market came to a halt while would-be buyers waited for the standards battle to be won by someone. Ultimately, that did not happen. Instead, the *International Telecommunication Union* (ITU) in its wisdom issued an "overlay standard" called V.90 that hid the incompatibilities of each standard from the other, making the choice of one over the other a non-issue.

The limitations of 56K modems stem from a number of factors. One of them is the fact that under current FCC regulations (specifically Part 68), line voltage supplied to a communications facility is limited such that the maximum achievable bandwidth is 53 Kbps in the downstream direction. Another limitation is that these devices require that only a single analog-to-digital conversion occur between the two end points of the circuit. This typically occurs on the downstream side of the circuit and usually at the interface point where the local loop leaves the central office. Consequently, downstream traffic is less susceptible to the quantization error ("digital noise") created during the analog-to-digital conversion process, while the upstream channel is affected by it and is therefore limited in terms of the maximum bandwidth it can provide. In effect, 56K modems, in order to achieve their maximum bandwidth, require that one end of the connection, typically the ISP or host end, be digital.

The good news with regard to 56K modems is that even in situations where the 56-Kbps speed is not achievable, the modem will fall back to whatever maximum speed it can fulfill. Furthermore, no premises' wiring changes are required, and since this device is really nothing more than a faster analog modem at the custmer site, the average customer is comfortable with migration to the new technology. This is certainly demonstrated by sales volume. As was mentioned before, most PCs today are automatically shipped with a 56K modem.

ISDN

The *Integrated Services Digital Network* (ISDN) has been the proverbial technological roller coaster since its arrival as a concept in the late 1960s. Internationally, it has enjoyed significant success as a true, digital local loop technology. In the United States, however, because of competing, incompatible hardware implementations, high cost, and marginal availability, its deployment has been spotty at best. In market areas where providers have made it available at reasonable prices, it has been quite successful. The good news is that since 1997 ISDN has experienced something of a phoenix-like resurgence that has occurred in lock-step with the growing demand for bandwidth. This growth, illustrated in Table 1-1,[5] will ultimately slow as other technologies such as xDSL and cable modems begin to disrupt the curve.

THE TECHNOLOGY

The typical non-ISDN local loop is analog. Voice traffic is carried from an analog telephone to the central office using a frequency-modulated carrier. Once at the *central office* (CO), the signal is usually digitized for transport within the digital network cloud. On the one hand, this is good because it means that there is a digital component to the overall transmission path. On the other hand, the loop is still analog, and as a result, the true promise of an end-to-end digital circuit cannot be realized.

In ISDN implementations, local switch interfaces must be modified to support a digital local loop. Instead of using analog frequency modulation to represent voice or data traffic carried over the local loop, ISDN digitizes the traffic at the origination point, either in the voice set itself or in an adjunct device known

TABLE 1-1   ISDN Growth

YEAR	BRI	PRI	TOTAL
1997	933K (+30%)	49K (+60%)	982K (+29%)
1998	1.25M (+35%)	79K (+60%)	1.229M (+36%)
1999	1.637M (+30%)	119K (+50%)	1.757M (+31%)
2000	2.047M (+25%)	167K (+40%)	2.214M (+26%)
2001	2.456M (+20%)	217K (+30%)	2.674M (+21%)
2002	2.824M (+15%)	261K (+15%)	3.086M (+15%)

[5]*Telephony Magazine*, July 1998.

as a *terminal adapter* (TA). The digital local loop then uses time division multiplexing to create multiple channels over which the digital information is transported, which provides a wide variety of truly integrated services.

THE BASIC RATE INTERFACE (BRI)

Two well-known implementations of ISDN exist. The most common (and the one intended primarily for residence and small business applications) is called the *Basic Rate Interface* (BRI). In BRI, the two-wire local loop supports a pair of 64-Kbps digital channels known as *B-Channels* as well as a 16-Kbps *D-Channel*, which is primarily used for signaling but can also be used by the customer for low-speed (up to 9.6 Kbps) packet data. The *B-Channels* can be used for voice and data, and in some implementations can be bonded together to create a single 128-Kbps channel for videoconferencing or other higher bandwidth applications.

Figure 1-4 shows the layout of a typical ISDN BRI implementation, including the alphabetic reference points that identify the regions of the circuit and the generic devices that make up the BRI. In this diagram, the LE is the local exchange or switch. The NT1 is the network termination device that serves as the demarcation point between the customer and the service provider. Among other things, the NT1 converts the two-wire local loop to a four-wire interface on the customer's premises. The *terminal equipment type 1* (TE1) is an ISDN-capable device, such as an ISDN telephone, while the *terminal equipment type 2* (TE2) is a non-ISDN-capable device, such as a POTS telephone. In the event that a TE2 is used, a TA must be inserted between the TE2

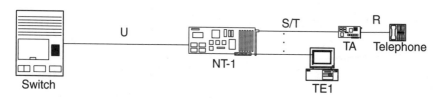

FIGURE 1-4   The ISDN BRI

and the NT1 to perform analog-to-digital conversions and rate adaptation. The TA that attaches a PC to an ISDN is analagous to the modem that connects the same PC to the PSTN.

The reference points mentioned earlier identify circuit components between the functional devices just described. The U reference point is the local loop, the S/T reference point sits between the NT1 and the TE1, and the R reference point is found between the TA and the TE2. ISDN protocols specifically address the S/T and U reference points.

BRI APPLICATIONS

Although BRI does not offer the stunning bandwidth that other more recent technologies do, its bondable 64-Kbps channels provide reasonable capacity for many applications. The two most common today are remote LAN and Internet access. For the typical remote worker, the bandwidth available through BRI is more than adequate, and new video compression technology even puts reasonable quality videoconferencing within the grasp of the end user at an affordable price. 64 Kbps makes short work of LAN-based text file downloads and reduces the time required for graphics-intensive Web page downloads to reasonable levels. The "World Wide Wait" becomes a minor annoyance in the grand scheme of things.

AO/DI

A relatively new arrival on the ISDN scene is an offering known as *Always On, Dynamic ISDN* (AO/DI). In AO/DI, the 16-Kbps D-Channel on the BRI, normally used for signaling, is recruited as a primary transport conduit for user data that normally employs a 64-Kbps B-Channel. If a user with AO/DI is connected to an e-mail service, for example, the packet-mode D-Channel can be used to deliver messages from the e-mail server to the user without requiring the services of a B-Channel. Should higher bandwidth be required to download a large attachment, view a Web page, or engage a more bandwidth-hungry application, a B-Channel can be activated, used, and torn down when it is no longer needed (see Figure 1-5). This technique is easier on the

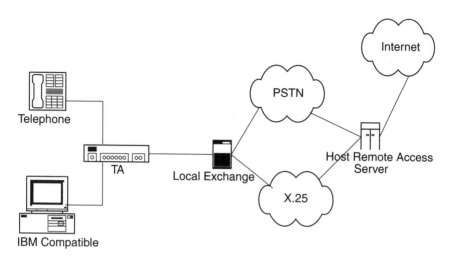

FIGURE 1-5   AO/DI

network because it makes more intelligent use of available bandwidth and therefore places less of a burden on the switch. It does, of course, require the customer to have an account with a packet services provider to handle the D-Channel packet traffic, but because these services are generally billed on a "volume of packets transmitted" basis, the cost is relatively low. Most users who employ AO/DI recognize that the cost of the packet service is far outweighed by the instantaneous access to online services that the technology provides.

THE PRIMARY RATE INTERFACE (PRI)

The other major implementation of ISDN is called the *Primary Rate Interface* (PRI). The PRI is really nothing more than a standard T-Carrier[6] in that it is a four-wire circuit (local loop), uses

---

[6]A T-Carrier, often called a T-1, is a 1.544-Mbps circuit that provides a total of 24 channels to the user, each of them 64 Kbps and known as a DS-0. Originally designed as an intra-office trunking scheme, it was believed that there would never be a reason why a customer would ever have to know of the existence of the T-Carrier, given the "vast" amounts of bandwidth it proffered. Today the T-Carrier is routinely deployed as a medium-bandwidth loop solution to the customer business premise for both voice and data applications.

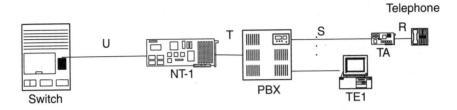

FIGURE 1-6 The ISDN PRI

AMI and B8ZS for density control and signaling, and sports a complement of 24 64-Kbps channels that can be distributed among a collection of users as the customer sees fit (see Figure 1-6). In PRI, the signaling channel operates at 64 Kbps (unlike the 16-Kbps D-Channel in the BRI) and is not accessible by the user. It is used solely for signaling purposes; it cannot carry user data. The primary reason for this is service protection. In the PRI, the D-Channel is used to control the goings-on of 23 B-Channels and therefore requires significantly more bandwidth than the BRI D-Channel. Furthermore, the PRI standards enable multiple PRIs to share a single D-Channel, which makes the D-Channel's operational consistency all the more critical.

The functional devices and reference points are not appreciably different from those of the BRI. Here the local loop is a four-wire T-Carrier rather than a traditional two-wire local loop, but it is still identified as the U reference point. In addition to an NT1, we now add an NT2, which is a service distribution device, usually a *Private Branch Exchange* (PBX), that allocates the PRI's 24 channels to customers. The S/T reference point is now divided. The S reference point sits between the NT2 and TEs, while the T reference point is found between the NT1 and the NT2.

PRI service also has the capability to "cluster" B-Channels into super-rate channels to satisfy the bandwidth requirements of higher bit rate services such as medical imaging and videoconferencing. These clusters of B-Channels are called *H-Channels* and are provisioned as shown in Table 1-2.

TABLE 1-2 H-Channels

CHANNEL	BANDWIDTH
H0	384 Kbps (6B)
H10	1.472 Mbps (23B)
H11	1.536 Mbps (24B)
H12	1.920 Mbps (30B)

PBX APPLICATIONS

The PRI's marketplace is the business community, and its primary advantage is pair gain, that is, the fact that it conserves copper pairs by multiplexing the traffic from multiple user channels onto a shared, four-wire circuit. Inasmuch as a PRI can deliver the equivalent of a minimum of 23 voice channels to a location over a single circuit, it is an ideal technology for a number of applications, including the interconnection of a PBX to a local switch, a dynamic bandwidth allocation for higher-end videoconferencing applications, and an interconnection between an ISP's network and that of the local telephone company.

Some PBXs are ISDN-capable on the line (customer) side, meaning that they have the capability to deliver ISDN services to users that emulate the services that would be provided over a direct connection to an ISDN-provisioned local loop. On the trunk (switch) side, the PBX is connected to the local switch via one or more T1s, which in turn provide access to the telephone network. This arrangement results in significant savings, faster call setups, more flexible administration of trunk resources, and the ability to offer a diversity of services through the granular allocation of bandwidth as required.

VIDEOCONFERENCING

Videoconferencing, an application that now enjoys widespread acceptance in the market thanks to affordable, effective video

CODECs, has moved from the boardroom to the home office. Although BRI provides adequate bandwidth for casual conferencing on a reduced-size PC screen, PRI is required for high-quality, television-scale sessions. Its capability to provision bandwidth on demand for such applications makes it an ideal solution. All major videoconferencing equipment manufacturers have embraced PRI as an effective connectivity solution for their products and have consequently designed their products in accordance with accepted international standards, specifically H.261 and H.320.

AUTOMATIC CALL DISTRIBUTION (ACD)

Another significant application for PRI is call routing in the call center environment. Ultimately, call centers represent critical decision points within a corporate environment, and the degree to which they are successful at what they do translates directly into visible manifestations of customer service. If calls are routed quickly and accurately based on some well-designed internal decision-making process, customers are happy, and the call center provides an effective image of the corporation and its services before the customer.

Automatic Call Distribution (ACD) is a popular switch feature that enables a customer to create custom "routing tables" in the switch so that incoming calls can be handled most effectively on a call-by-call basis. The ACD feature often relies on information culled from both corporate sources and the SS7 network's SCP databases, but the ultimate goal is to provide what appears to each caller to be customized answering. For example, some large call centers handle incoming calls from a variety of countries and therefore potentially different language groups. Using caller ID information, an ACD can filter calls as they arrive and route them to language-appropriate operators. Similarly, the ACD can change the assignment of voice channels on a demand basis. For example, if the call center is large and receives calls from multiple time zones simultaneously, there may be a need to add incoming trunks to one region of the call center based on time-of-day call volumes, while reducing the total coverage in another. By reallocating the 64K channels of the PRI(s), bandwidth can be made available as required. Furthermore, it can be done automatically, triggered by the time of day or some other pre-identified event.

IN SUPPORT OF THE REMOTE AGENT

Many companies now support distributed call centers. In this model, agents are not necessarily physically collocated, but may in fact be working in their homes. Each remote agent has a BRI line to their location that provides them with two channels, one for voice, the other for access to database information and whatever applications they must use to perform their jobs. The ACD feature automatically routes calls to each agent according to some predefined set of functions. The incoming calls appear on one B-channel; the other is reserved for database access.

This model has been applauded by both corporations and government agencies alike, because it supports the basic tenets of telecommuting. Employees who work this way are happy because they can stay home to care for children or elderly parents and are generally (according to most studies that have been conducted) much more productive than those who work in a traditional office setting. The government is happy as well, because telecommuting reduces the total number of cars on the road, thus supporting federal clean air mandates. Corporations benefit because real estate costs can be reduced. If they legitimately support the program, they can provision less space for employees and implement "hoteling" philosophies. Hoteling is a relatively new technique for office space management in which employees, particularly those who travel for a significant proportion of their professional time, do not have dedicated cubicle or office space. Instead, their personal effects are stored in rollaround cabinets that can easily be moved from one place to another. On those occasions when a "road warrior" employee must work in the office, they are dynamically assigned a cubicle, to which they trundle their personal cabinet. Once there, they call a number that causes the switch to automatically assign their office phone number to the phone line in that cubicle.

Ironically, ISDN is the earliest example of true technological convergence. The value of ISDN does not lie in the fact that it enables multiple information streams to share a common physical transmission channel. That capability has been available since the turn of the century when scientists first invented harmonic telegraphs. Its value lies in the fact that it enables a *logical integration of services* over a shared channel, the essence of the convergence phenomenon. And although it does not offer the high bandwidth provided by competitive wireline technologies such as cable modems and xDSL, it has the advantage of being able to offer a full suite of capabilities, including reasonable bandwidth for quite a few applications *right now*, a claim that alternative technologies can't necessarily make.

T-CARRIER AND VOICE DIGITIZATION

The voice network, including both transmission facilities and switching components, was exclusively analog in nature until 1962, when T-Carrier emerged as an intra-office trunking scheme. The technology was originally introduced as a short-haul, four-wire facility to serve metropolitan areas. Over the years, it evolved to include coaxial cable facilities, digital microwave systems, fiber, and satellite, and today is a commonly deployed access technology for medium-speed applications. Amazingly, when it was first introduced, there was no concept that "the outside world" (that is, outside the walls of the central office) would ever have to be aware of its existence. After all, a customer would never require the almost unimaginable bandwidth that T-Carrier provided!

As the network topology improved, so too did the switching infrastructure. In 1976, AT&T introduced the 4ESS switch primarily for toll applications and followed it up with the 5ESS in 1981 for local switching access as well as a variety of remote switching capabilities. Nortel, Siemens, and Ericsson all followed suit with equally capable hardware. In 1983, the first tariff for T-1 was released, and the service was on its way to becoming mainstream.[7]

The goal of digitizing the human voice for transport across an all-digital network grew out of work performed at Bell Laboratories shortly after the turn of the century. That work led to a discrete understanding of not only the biological nature and spectral makeup of the human voice, but also to a better understanding of language, sound patterns, and the sounded emphases that comprise spoken language.

THE NATURE OF VOICE

A typical voice signal comprises frequencies that range from approximately 30 Hz to 20 KHz. Most of the speech energy, however, lies between 300 and 3,300 Hz, the so-called voice band. Early experiments showed that the frequencies below 1 KHz provide the bulk of recognizability and intelligibility, while the higher frequencies provide richness, articulation, and natural sound to the transmitted signal.

As we noted earlier, the human voice comprises a remarkably rich mix of frequencies, and this richness comes at a considerable price. In order for telephone networks to transmit voice's entire spectrum of frequencies, significant network bandwidth must be made available to every ongoing conversation. There is a substantial price tag attached to bandwidth. It is a finite commodity within the network, and the more of it that is consumed, the more it costs.

Thankfully, work performed at Bell Laboratories at the beginning of the 20th century helped network designers confront this challenge head-on. When the telephone network first began to spread its tentacles across the continent, there were no switches. Initially, subscribers in towns bought individual

---

[7]It seems that there are as many explanations for the "T" in T-Carrier as there are circuit miles of the service deployed. Let us dispense with the rumors: the T stands for "terrestrial."

phones and phone lines to each person they had a need to speak with, resulting in the famous photographs of metropolitan telephone poles with tier upon tier of cross-pieces, festooned with aerial wire, and its mathematical representation: $n(n-1)/2$, the total number of circuits that would be required for n people to be able to speak with everyone else in the community. For example, if 100 people in a neighborhood wanted to be able to speak with each other, they would require the phone company to install a total of $(100)*(99)/2$, or 4,950 circuits for 100 people. Clearly, this was economically out of the question, not to mention the fact that the quantity of aerial cable threatened to block out the sun, potentially precipitating the next ice age and the end of civilization as we know it.

The solution to this quandary came in two forms: central office switches and multiplexing. Central office switches enabled the phone company to provide the same level of connectivity, but required each customer to have only *one* circuit. The responsibility for setting up the connection to the called party resided with the switch in the central office, rather than with the customer. The first "switches," of course, were operators ("Sarah, get me Andy over at the courthouse!"). True mechanical switches didn't arrive until 1892 when Almon Strowger's Step-by-Step switch was first installed by Automatic Electric.

Equally important was the concept of multiplexing, which enabled multiple conversations to be carried simultaneously across a single shared physical circuit. The first systems used *frequency-division multiplexing* (FDM), a technique made possible by the development of the vacuum tube in which the range of available frequencies is divided into "chunks" that are then parceled out to subscribers. For example (and this is *only* an example), subscriber #1 might be assigned the range of frequencies between 0 and 4,000 Hz, while subscriber #2 is assigned 4,000 to 8,000 Hz, #3 8,000 to 12,000 Hz, and so on, up to the maximum range of frequencies available in the system. In frequency-division multiplexing, we often observe that users are given "some of the frequency all of the time," meaning that they are free to use their assigned frequency allocation at any time but may *not* step outside the bounds. Early FDM systems were capable of transporting 24-4 KHz channels for an overall system bandwidth of 96 KHz. FDM, while largely replaced today by more efficient systems that will be discussed later, is still used in cellular telephone and microwave systems among others.

FDM worked well in the early telephone systems. Because the lower regions of the 300- to 3,300-Hz voice band carry the frequency components that enable recognizability and intelligibility, telephone engineers concluded that while the higher frequencies enrich the transmitted voice, they are not necessary for both parties to recognize and understand each other. This understanding of the makeup of the human voice helped them create a network that is capable of faithfully reproducing the sounds of a conversation while keeping the cost of consumed bandwidth to a minimum. Instead of assigning the full complement of 20 KHz to each end of a conversation, they employed filters to bandwidth-limit each user to approximately 4,000 Hz, a resource savings of some 80 percent. Within the network, subscribers were FDMed across shared physical facilities, thus allowing the telephone company to efficiently conserve network bandwidth.

As with any technology, there were downsides to FDM. It is an analog technology and therefore suffers from the shortcomings that have historically plagued all transmission systems. The wire that information is transmitted over behaves like a long-wire antenna, picking up noise along the length of the transmission path and effectively homogenizing it with the voice signal. Additionally, the power of the transmitted signal diminishes over distance, and if the distance is great enough, the signal will have to be amplified to make it intelligible at the receiving end. Unfortunately, the amplifiers used in the network are not particularly discriminating. They have no way of separating the voice wheat from the noise chaff, as it were. The result is that they convert a weak, noisy signal into a loud noisy signal. Better, but far from ideal. A better solution was needed.

The better solution came about with the development of *time-division multiplexing* (TDM), which became possible with the development of the transistor and integrated circuit electronics. TDM is a digital transmission scheme, which implies a small number of discrete signal states rather than the essentially infinite range of values employed in analog systems. Although digital systems are as susceptible to noise impairment as their analog counterparts, the discrete nature of their binary signaling makes it relatively easy to separate the noise from the transmitted signal. In T-Carrier systems, only three valid signal values exist: one positive, one negative, and zero. Anything else is construed to be noise. It is therefore a trivial exercise for digital repeaters (the digital equivalent of an analog amplifier) to discern what is desirable and what is not, thus eliminating the problem of cumulative noise.

Because of their rich frequency content, digital signals require a broad transmission channel. They cannot be transmitted across the bandwidth-limited channels of the traditional telephone network. In digital T-Carrier facilities, the equipment that restricts the individual transmission channels to four-KHz chunks is removed, thus giving each user access to the full breadth of available spectrum across the shared physical medium. In FDM systems, we observed that we give users "some of the frequency all of the time." In TDM systems, we turn that around and give users "*all* of the frequency *some* of the time" (see Figure 1-7).

Digitization brings with it a cadre of advantages, including improved voice and data transmission quality, better maintenance and troubleshooting capabilities (and therefore reliability), and dramatic improvements in configuration flexibility. In T-Carrier systems, the TDM is known as a channel bank. Under normal circumstances, it enables 24 circuits to share a single, four-wire facility, as shown in Figure 1-8.

VOICE DIGITIZATION

The process of converting analog voice to a digital representation in the modern network is a logical and straightforward process. It comprises four distinct steps: *Pulse Amplitude Modulation* (PAM) sampling, in which the amplitude of the

Customer #1:	0 – 4 KHz
Customer #2:	4 – 8 KHz
Customer #3:	8 – 12 KHz
Customer #4:	12 – 16 KHz
Customer #5:	16 – 20 KHz

Frequency Division Multiplexing

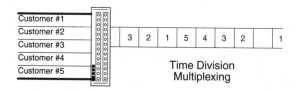

Time Division Multiplexing

FIGURE 1-7   FDM versus TDM

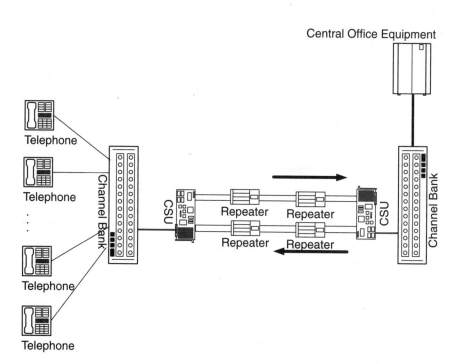

FIGURE 1-8   Typical T-1 span layout

incoming analog waveform is sampled once every 125 microseconds; companding, during which the values are weighted toward those most receptive to the human ear; quantization, in which the weighted samples are given values on a nonlinear scale; and finally encoding, in which each value is assigned a distinct binary value. Each of these stages of *Pulse Code Modulation* (PCM) will be discussed in detail.

PULSE CODE MODULATION (PCM)

Thanks to work performed by Harry Nyquist at Bell Laboratories in the 1920s, we know that to optimally represent an analog signal as a digitally encoded bitstream, the analog signal must be sampled at a rate that is equal to twice the highest frequency of the channel over which the signal is to be transmitted. Since each analog voice channel is allocated four KHz of bandwidth, it follows that each voice signal must be sampled at twice that rate, or 8,000 samples per second. In fact, that is precisely what happens in T-Carrier systems. The standard T-Carrier multiplexer accepts inputs from 24 analog channels. Each channel is sampled in turn, every one eight-thousandth of a second in round-robin fashion, resulting in the generation of 8,000 pulse amplitude samples from each channel every second.[8] This PAM process represents the first stage of PCM, the process by which an analog signal is converted to a digital signal for transmission across the T-Carrier network.

The second stage of PCM is called quantization. In quantization, we assign values to each sample within a constrained range. For illustration purposes, imagine what we now have before us. We have "replaced" the continuous analog waveform of the signal with a series of amplitude samples that are close enough together that we can discern the shape of the original wave from their collective amplitudes. Imagine also that we

---

[8]The sampling rate is important for several reasons. If the sampling rate is too high, too much information is transmitted, and bandwidth is wasted. If the sampling rate is too low, then we run the risk of aliasing. *Aliasing* is the interpretation of the sample points as a false waveform due to the paucity of samples.

have graphed these samples in such a way that the "wave" of sample points meanders above and below an established zero point on the *x*-axis, so that some of the samples have positive values and others are negative, as shown in Figure 1-9.

The amplitude levels enable us to assign values to each of the PAM samples, although a glaring problem with this technique should be obvious to the careful reader. Very few of the samples actually line up exactly with the amplitudes delineated by the graphing process. In fact, most of them fall *between* the values. It doesn't take much of an intuitive leap to see that several of the

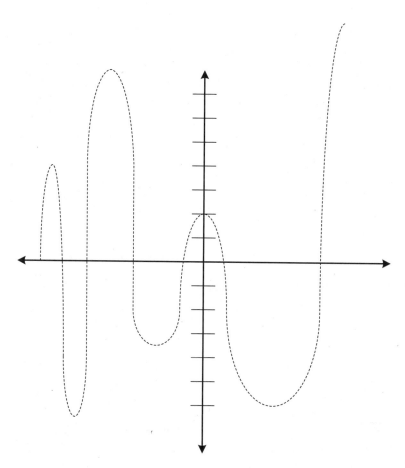

FIGURE 1-9   Sampling a voice signal

different samples will be assigned the same digital value by the coder-decoder that performs this function, yet they are clearly *not* the same amplitude. This inaccuracy in the measurement method results in a problem known as *quantizing noise* and is inevitable when linear measurement systems, such as the one suggested by the drawing, are employed in *coder-decoders* (CODECs).

Needless to say, design engineers recognized this problem rather quickly and equally quickly came up with an adequate solution. It is a fairly well-known fact among psycholinguists and speech therapists that the human ear is far more sensitive to discrete changes in amplitude at low volume levels than it is at high volume levels, a fact not missed by the network designers that optimize the performance of digital carrier systems intended for voice transport. Instead of using a linear scale for digitally encoding the PAM samples, they designed and employed a non-linear scale that is weighted with much more granularity at low volume levels—that is, close to the zero line —than at the higher amplitude levels, as shown in Figure 1-10. In other words, the values are extremely close together near the *x*-axis and get farther and farther apart as they travel up and

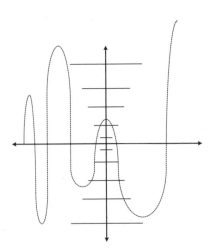

FIGURE 1-10   Companding the voice signal

down the y-axis. This non-linear approach keeps the quantizing noise to a minimum at the low amplitude levels where hearing sensitivity is the highest and enables it to creep up at the higher amplitudes where the human ear is less sensitive to its presence. It turns out that this is not a problem because the inherent shortcomings of the mechanical equipment (microphones, speakers, and the circuit itself) introduce slight distortions at high amplitude levels that hide the effect of the non-linear quantizing scale.

This technique of "compressing" the values of the PAM samples to make them fit the non-linear quantizing scale results in a bandwidth savings of more than 30 percent. In fact, the actual process is called *companding*, because the sample is first compressed for transmission and then expanded for reception at the far end, hence the term.

The actual graph scale is divided into 255 distinct values above and below the zero line.[9] Eight segments exist above the line and eight below (one of which is the shared zero point). Each segment, in turn, is subdivided into 16 steps. A bit of binary mathematics now enables us to convert the quantized amplitude samples into an eight-bit value for transmission. For the sake of demonstration, let's consider a positive sample that falls into the eleventh step in segment seven. The conversion would take on the following representation:

0 111 1011

where the initial 0 indicates a positive sample, 111 indicates the seventh segment, and 1011 indicates the eleventh step in the segment. We now have an eight-bit representation of an analog amplitude sample that can be transmitted across a digital network and then reconstructed with its many counterparts as an accurate representation of the original analog waveform at the receiving end. This entire process is known as PCM, and the result of its efforts is often referred to as "toll-quality voice."

---

[9]In the U.S. and a handful of other countries including Canada and Japan, the encoding scheme is known as μ-Law (Mu-Law). The rest of the world relies on a slightly different standard known as A-Law.

ALTERNATIVE DIGITIZATION TECHNIQUES

Although PCM is perhaps the best-known, high-quality voice digitization process, it is by no means the only one. Advances in *Complementary Metal Oxide Semiconductor* (CMOS) architecture and improvements in the overall quality of the telephone network have made it possible for encoding schemes to be developed that use far less bandwidth than traditional PCM. For the corporation, this means a significant savings in bandwidth and therefore cost. In this next section, we will consider some of these techniques.

Adaptive Differential Pulse Code Modulation (ADPCM)

Adaptive Differential Pulse Code Modulation (ADPCM) is a technique that enables toll-quality voice signals to be encoded at halfrate (32 Kbps) for transmission. ADPCM relies on the predictability that is inherent in human speech to reduce the amount of information required. The technique still relies on PCM encoding, but adds an additional step to carry out its task. The 64-Kbps PCM-encoded signal is fed into an ADPCM transcoder, which considers the *prior* behavior of the incoming stream to create a prediction of the behavior of the *next* sample. Here's where the magic happens. Instead of transmitting the actual value of the predicted sample, it encodes in four bits and transmits the *difference* between the actual and predicted samples. Since the difference from sample to sample is typically quite small, the results are generally considered to be close to toll quality. This four-bit transcoding process, which is based on the known behavior characteristics of human voice, enables the system to transmit 8,000 four-bit samples per second, thus reducing the overall bandwidth requirement from 64 Kbps to 32 Kbps.[10]

---

[10]It should be noted that ADPCM works well for voice because the encoding and predictive algorithms are based upon its behavior characteristics. It does not, however, work as well for high bit rate data (above 4,800 bps), which has an entirely different set of behavior characteristics.

Continuously Variable Slope Delta Modulation (CVSD)

Continuously Variable Slope Delta Modulation (CVSD) is a unique form of voice encoding that relies on the individual values of individual bits to predict the behavior of the incoming signal. Instead of transmitting the volume (height or *y*-value) of PAM samples, CVSD transmits information that measures the changing slope of the waveform. So, instead of transmitting the actual change itself, it transmits the *rate* of change.

To perform its task, CVSD uses a reference voltage to which it compares all incoming values. If the incoming signal value is less than the reference voltage, then the CVSD encoder reduces the slope of the curve to make its approximation better mirror the slope of the actual signal. If the incoming value is *more* than the reference value, then the encoder will increase the slope of the output signal, again causing it to approach and therefore mirror the slope of the actual signal. With each recurring sample and comparison, the step function can be increased or decreased as required. For example, if the signal is increasing rapidly, then the steps are increased one after the other in a form of step function by the encoding algorithm. Obviously, the reproduced signal is not a particularly exact representation of the input signal. In practice, it is pretty jagged. Filters therefore are used to smooth the transitions.

CVSD is typically implemented at 32 Kbps, although it can be implemented at rates as low as 9,600 bps. At 16 to 24 Kbps, recognizability is still possible. Below 9,600, recognizability is seriously affected, although intelligibility is not.

Linear Predictive Coding (LPC)

We mention *Linear Predictive Coding* (LPC) here only because it has carved out a niche for itself in certain voice-related applications such as voice mail systems, automobiles, aviation, and electronic games. LPC is a complex process, implemented completely in silicon, which enables voice to be encoded at rates as low as 2,400 bps. The result is far from toll-quality but is certainly intelligible, and its low bit rate capability gives it a distinct advantage over other systems.

LPC relies on the fact that each sound created by the human voice has unique attributes, such as frequency range, resonance, and loudness, among others. When voice samples are created in LPC, these attributes are used to generate *prediction coefficients*. These predictive coefficients represent linear combinations of previous samples, hence the name *Linear Predictive Coding*.

Prediction coefficients are created by taking advantage of the known *formants* of speech, which are the resonant characteristics of the mouth and throat, which give speech its characteristic timbre and sound. This sound, referred to by speech pathologists as the buzz, can be described by both its pitch and its intensity. LPC therefore models the behavior of the vocal cords and the vocal tract itself.

To create the digitized voice samples, the buzz is passed through an inverse filter that is selected based upon the value of the coefficients. The remaining signal, after the buzz has been removed, is called the *residue*.

In the most common form of LPC, the residue is encoded as either a *voiced* or *unvoiced* sound. Voiced sounds are those that require vocal cord vibration, such as the g in glare, the b in boy, or the d and g in dog. Unvoiced sounds require no vocal cord vibration, such as the h in how, the sh in shoe, or the f in frog. The transmitter creates and sends the prediction coefficients, which include measures of pitch, intensity, and whatever voiced and unvoiced coefficients are required. The receiver undoes the process. It converts the voice residue, pitch, and intensity coefficients into a representation of the source signal, using a filter similar to the one used by the transmitter to synthesize the original signal.

DIGITAL SPEECH INTERPOLATION (DSI)

Human speech has many measurable (and therefore predictable) characteristics, one of which is a tendency to have embedded pauses. As a rule, people do not spew out a series of uninterrupted sounds. They tend to pause for emphasis, to collect their thoughts, or to reword a phrase while the other person listens quietly on the end of the line. When speech technicians monitor these pauses, they discover that during considerably more than half of the total connect time, the line is silent.

Digital speech interpolation (DSI) takes advantage of this characteristic silence to drastically reduce the bandwidth required for a single channel. Whereas 24 channels can be transported over a typical T-1 facility, DSI enables as many as 120 conversations to be carried over the same circuit. The format is proprietary and requires the setting aside of a certain amount of bandwidth for overhead.

A form of statistical multiplexing lies at the heart of DSI's functionality. Standard T-Carrier is a TDM scheme, in which channel ownership is assured. A user assigned to channel three will *always* own channel three, regardless of whether they are actually using the line. In DSI, channels are not owned. Instead, large numbers of users share a pool of available channels. When a user starts to talk, the DSI system assigns an available timeslot to that user and notifies the receiving end of the assignment. This system works well when the number of users is large, because statistical probabilities are more accurate and indicative of behavior in larger populations than in smaller ones.

DSI has a downside, of course, and it comes in several forms. *Competitive clipping* occurs when people outnumber the available channels, resulting in someone being unable to talk. *Connection clipping* occurs when the receiving end fails to learn what channel a conversation has been assigned within a reasonable amount of time, resulting in signal loss. Two approaches have been created to address these problems. In the case of competitive clipping, the system intentionally "clips" off the front end of the initial word of the second person who speaks. This technique is not optimal but does prevent loss of the conversation and also obviates the problem of clipping out the middle of a conversation, which would be more difficult for the speakers to recover from. The loss of an initial syllable or two can be mentally reconstructed far more easily than the middle of a sentence.

A second technique used to recover from clipping problems is to temporarily reduce the encoding rate. The typical encoding rate for DSI is 32 Kbps. In certain situations, the encoding rate may be reduced to 24 Kbps, thus freeing up bandwidth for additional channels. Both techniques are widely utilized in DSI systems.

FRAMING AND FORMATTING IN T-1

The standard T-Carrier multiplexer accepts inputs from 24 sources, converts the inputs to PCM bytes, then TDMs the samples over a shared 4-wire facility. Each of the 24 input channels yields an 8-bit sample, in round-robin fashion, once every 125 microseconds (8,000 times per second). This yields an overall bit rate of 64 Kbps for each channel (8 bits per sample $\times$ 8,000 samples per second). The multiplexer gathers one 8-bit sample from each of the 24 channels and aggregates them into a 192-bit frame. To the frame it adds a frame bit, which expands the frame to a 193-bit entity. The frame bit is used for a variety of purposes that will be discussed in a moment.

The 193-bit frames of data are transmitted across the 4-wire facility at the standard rate of 8,000 frames per second for an overall T-1 bit rate of 1.544 Mbps. Keep in mind that 8 Kbps of the bandwidth consist of frame bits; only 1.536 Mbps belong to the user.

The earliest T-Carrier equipment was referred to as a D1 channel bank and was considerably more rudimentary in function than modern systems. In D1, every eight-bit sample carried seven bits of user information (bits one through seven) and one bit for signaling (bit eight). The signaling bits were used for exactly that, indications of the status of the line (on-hook, off-hook, busy, high and dry, and so on), while the seven user bits carried encoded voice information. Since only seven of the eight bits were available to the user, the result was considered to be less than toll quality (128 possible values, rather than 256). The frame bits, which in modern systems indicate the beginning of the next 192-bit frame of data, toggled back and forth between zero and one.

As time went on and the stability of network components improved, an improvement on D1 was sought after and found. Several options were developed, but the winner emerged in the form of the D4 or superframe format. Rather than treat a single 193-bit frame as the transmission entity, superframe gangs together 12 193-bit frames into a 2,316-bit entity, which obviously includes 12 frame bits. Please note that the bit rate has not changed; we have simply changed our view of what constitutes a frame.

Since we now have a single (albeit large) frame, we clearly don't need 12 frame bits to frame it. Consequently, some of them can be redeployed for other functions. In superframe, six of the odd-numbered frame bits are referred to as terminal framing bits and are used to synchronize the channel bank equipment. The other six framing bits are called signal framing bits and indicate to the receiving device *where* robbed-bit signaling occurs.

In D1, the system reserves one bit from every sample for its own signaling purposes, which succeeds in reducing the user's overall throughout. In D4, that is no longer necessary. Instead, we signal less frequently and only occasionally rob a bit from the user. In fact, because the system operates at a high transmission speed, network designers determine that signaling can occur relatively infrequently and still convey adequate information to the network. Consequently, bits are robbed from the sixth and eighth iteration of each channel's samples, and then only the least significant bit from each sample. The resulting change in voice quality is negligible.

Back to the signal framing bits. Within a transmitted superframe, the second and fourth signal framing bits are the same, but the sixth toggles to the opposite value, indicating to the receiving equipment that the samples in that subframe of the superframe should be checked for signaling state changes. The eighth and tenth signal framing bits stay the same as the sixth but toggle back to the opposite value once again in the twelfth, indicating once again that the samples in that subframe should be checked for signaling state changes.

Although superframe continues to be widely utilized, an improvement came about in the 1980s in the form of *extended superframe* (ESF). It groups 24 frames into an entity instead of twelve and, like superframe, reuses some of the frame bits for other purposes. Bits 4, 8, 12, 16, 20, and 24 are used for framing and form a constantly repeating pattern (001011 . . .). Bits 2, 6, 10, 14, 18, and 22 are used as a 6-bit *cyclic redundancy check* (CRC) to check for bit errors on the facility. Finally, the remaining bits—all of the odd frame bits in the frame—are used as a 4-Kbps facility data link for end-to-end diagnostics and network management tasks. In this regard, ESF provides a major benefit over its predecessors. In earlier systems, if a customer reported trouble on the span, the span would have to be taken out of service for testing. With ESF, that is no longer necessary because of the added functionality provided by the CRC and the facility data link. The 4-Kbps *embedded operations channel* (EOC) that results from the efficient reuse of framing bits in ESF provides a dedicated facility between the *channel service units* (CSUs) on a T-Carrier circuit over which the service provider can transmit diagnostic signals without affecting the customer.

The T-1 framing and transmission standard is used in North America and Japan, while the remainder of the world uses what is known as the CEPT E-1 standard. CEPT, the European Council on Post and Telecommunications Administrations, has largely been replaced today by the *European Telecommunications Standards Institute* (ETSI), but the name is still occasionally used.

E-1 differs from T-1 on several key points. First, it boasts a 2.048-Mbps rate, rather than the 1.544-Mbps facility found in T-1. Second, it utilizes a 32-channel frame rather than 24. Channel 1 contains framing information and a *4-bit cyclic redundancy check* (CRC-4), channel 16 contains all signaling information for the frame, and channels 1 through 15 and 17 through 31 transport user traffic.

A number of similarities exist between T-1 and E-1 as well. Their channels are all 64 Kbps and frames are transmitted 8,000 times per second. And although T-1 groups together 24 frames to create an extended superframe, E-1 groups 16 frames to create what is known as a *ETSI multiframe*. The multiframe is subdivided into two submultiframes and the CRC-4 in each one is used to check the integrity of the submultiframe that preceded it.

A final word about T-1 and E-1. Because T-1 is a departure from the international E-1 standard, it is incumbent upon the T-1 provider to perform all interconnection conversions between T-1 and E-1 systems. For example, if a call arrives in the U.S. from a European country, the receiving American carrier must convert the incoming E-1 signal to T-1. If a call originates from Canada and is terminated in Australia, the Canadian originating carrier must convert the call to E-1 before transmitting it to Australia.

CABLE-BASED ACCESS TECHNOLOGIES

In 1950, Ed Parsons placed an antenna on a hillside above his home in Washington, attached it to a coaxial cable distribution network, and began to offer television service to his friends and neighbors. Prior to his efforts, the residents of his town were unable to pick up broadcast channels because of the blocking effects of the surrounding mountains. Thanks to Parsons, *community antenna television* (CATV) was born. From its roots came cable television.

Since that time, the cable industry has grown into a $25 billion industry. In the U.S. alone, 10,000 headends deliver content to 60 million homes in more than 22,000 communities over more than a million miles of coaxial and fiber-optic cable. And as the industry's network has grown, so too have the aspirations of those deploying it. Their goal is to make it much more than a one-way medium for the delivery of television and pay-per-view. They want to provide a broad spectrum of interactive, two-way services that will enable them to compete head-to-head with the telephony industry. To a large degree, they are succeeding. The challenges they face, however, are daunting.

THE BROADBAND BALKANS

Unlike the telephone industry that began under the control and design direction of a small number of like-minded individuals (such as Alexander Bell and Theodore Vail, among others), the cable industry came about thanks to the combined efforts of hundreds of innovators, each of them building on Ed Parson's original idea. As a consequence, the industry, while enormous, was in many ways fragmented. Powerful industry leaders like TCI's John Malone and Time Warner's Gerald Levine were able to unite the many companies, turning a loosely cobbled-together collection of players into cohesive, powerful corporations with a shared vision of what they were capable of accomplishing.

Today the cable industry is a force to be reckoned with, and upgrades to the original network are underway. This is a crucial activity that will ensure the success of the industry's ambitious business plan and provide a competitive balance for the traditional telcos.

THE CABLE NETWORK

The traditional cable network is an analog system based on a tree-like architecture. The headend, which serves as the signal origination point, is connected to the downstream distribution network by a one-inch diameter rigid coaxial cable, as shown in Figure 1-11. That cable delivers the signal, usually a 450-MHz collection of 6-MHz channels, to a neighborhood where splitters divide the signal and send it down a half-inch-diameter semi-rigid coax that typically runs down a residential street. At each house, another splitter pulls off the signal and feeds it to the set-top box in the house over the drop wire, a "local loop" of flexible quarter-inch coaxial cable.

Although this architecture is perfectly adequate for the delivery of one-way television signals, its shortcomings for other services should be fairly obvious to the reader. First of all, it is, by design, a unidirectional broadcast system. It does not typically have the capability to support upstream traffic (from the customer toward the headend) and is therefore not suited for interactive applications. Second, because of its design, the network is prone to significant failures that have the potential to affect large numbers of customers. The tree structure, for example, means that if a failure occurs along any "branch" in the tree, every customer from that point downward loses service. Contrast this with the telephone network where customers

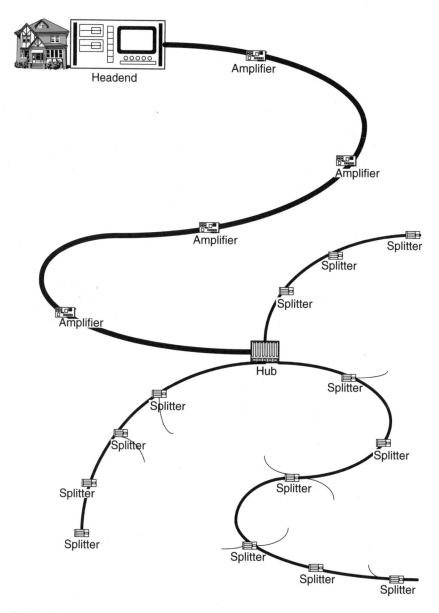

FIGURE 1-11   An analog cable network

have a dedicated local loop over which their service is delivered. Third, because the system is analog, it relies on amplifiers to keep the signal strong as it is propagated downstream. These amplifiers are powered locally; they do not have access to central office power as devices in the telephone network do. Consequently, a local power failure can bring down the network's capability to distribute service in that area.

The third issue is one of customer perception. For any number of reasons, it is generally perceived that the cable network is not as capable or as reliable as the telephone network. As a consequence of this perception, the cable industry is faced with the daunting challenge of convincing potential voice and data customers that they are in fact capable of delivering high-quality service. Some of the concerns are justified. In the first place, the telephone network has been in existence for almost 125 years, during which time its operators have learned how to optimally design, manage, and operate it in order to provide the best possible service. The cable industry, on the other hand, came about 50 years ago, and didn't benefit from the rigorously administered, centralized management philosophy that characterized the telephone industry. Additionally, the typical 450-MHz cable system did not have adequate bandwidth to support the bidirectional transport requirements of new services.

Furthermore, the architecture of the legacy cable network, with its distributed power delivery and tree-like distribution design, does not lend itself to the same high degree of redundancy and survivability that the telephone network offers. Consequently, cable providers have been hard-pressed to convert customers who have become vigorously protective of their telecommunications services.

NEXT GENERATION CABLE SYSTEMS

Faced with these harsh realities and the realization that the existing cable plant could not compete with the telephone network in its original analog incarnation, cable engineers began a major rework of the network in the early 1990s. Beginning with the headend and working their way outward, they progressively redesigned the network to the extent that in many areas of the country their coaxial local loop is capable of competing on equal footing with the telco's twisted pair, and in some cases besting it.

The process they used to reach this rather remarkable stage in their competitive technological development comprised four phases. In the first phase, they converted the content generation process, that is, the headend, from analog to digital. This enabled them to digitally compress the content, resulting in a far more efficient utilization of the available system bandwidth. Second, they undertook an ambitious physical upgrade of the coaxial plant, replacing the one-inch trunk and half-inch distribution cable with optical fiber. This brought about several desirable results. First, by using a fiber feeder infrastructure, network designers were able to eliminate a significant percentage of the amplifiers responsible for such a high percentage of the failures the network experienced due to power problems in the field. Second, the fiber made it possible to provision significantly more bandwidth than coaxial systems. Third, because the system was now largely digital, it suffered less from noise-related errors than its analog predecessor. Finally, an upstream return channel was provisioned, which made it possible to deliver true interactive services such as voice, Web surfing, and videoconferencing.

The third phase of the conversion had to do with the equipment provisioned at the user's premises. The analog set-top box has now been (or soon will be) replaced with a digital device that has the capability to take advantage of the digital nature of the network, including access to the upstream channel. It will decompress digital content, perform content ("stuff") separation, and provide the network interface point for data and voice devices.

The last phase of the overall process is business conversion. Cable providers look forward to the day when their networks will compete on equal footing with the twisted pair networks of the telephone company, and customers will see them as viable competitors. In order for this to happen, they must demonstrate that their network is capable of delivering a broad variety of competitive services, that the network is robust, that they have *operations support systems* (OSSs) that will guarantee the robustness of the infrastructure, and that they are cost-competitive with incumbent providers. TCI, for example, has taken significant steps to do this by joining forces with AT&T, a key member of the communications business that is now brand-ing TCI services with its own well-respected name. This will have a positive impact on TCI's ability to effectively penetrate the market, a process that will be discussed in detail in the section on company convergence.

HYBRID FIBER COAX SYSTEMS (HFC)

The current breed of cable delivery networks that shows the most promise is called a *Hybrid Fiber Coax* (HFC) system. As the name implies, HFC systems comprise a fiber feeder architecture that delivers broadband services to a served area. In the neighborhood or at the business location to which the service is being delivered, the fiber terminates at a hub. There the optical signal is converted to electrical and delivered to the premises on traditional coaxial cable.

This architecture has several advantages. First of all, the optical feeder can be provisioned on a dual fiber ring architecture, optimally running the *Synchronous Optical Network* (SONET) physical layer protocol. In most fiber ring deployments, one ring is designated as the primary path, while the other is provisioned as a backup. Among other things, SONET provides a feature known as *Automatic Protection Switching* (APS), which constantly monitors the integrity of the rings. If the primary ring fails for any reason (backhoe fade[11] being the most common), devices on the ring automatically switch over to the backup ring, limiting the outage to approximately 50 ms.

Another advantage is cost. Signals transported across optical fiber degrade far more slowly and accumulate less noise than those transported on copper or coaxial cable, resulting in the capability to build far longer spans with fewer repeaters—the digital equivalent of an amplifier. Furthermore, the bandwidth available in these systems is quite high, and as a consequence,

---

[11]A euphemism used in the transport industry to describe what happens when a backhoe severs a cable while digging. The fade, of course, is instantaneous and usually catastrophic.

more services can be provisioned to a broader population of customers.

Most cable operators have designated different ranges of the available system bandwidth for the transport of both digital and analog services. The digital domain will carry compressed video and audio, while the analog portion of the cable spectrum will potentially transport voice and data.

A word about cable telephony. In order to transport voice, cable systems must be interconnected at the headend to the PSTN. This means that local and long distance carriers must have *points of presence* (POPs) in cable provider headend locations so that calls can be handed off to the switched network. This process is already underway. The AT&T/TCI combination is a good example. AT&T will provision a POP in each TCI headend location, enabling easy cross-connections between the TCI local telephony network and the AT&T long-distance network.

CABLE MODEMS

As cable providers have progressively upgraded their networks to include more fiber in the backbone, their offer to provide two-way access to the Internet was a natural next step. Enter the cable modem. This new technology offers access speeds of up to 10 Mbps, and although availability is still spotty, studies indicate that only about 25 percent of the country's cable subscribers can actually get the service. The number is growing as cable providers respond to growing demand for access bandwidth. Today the cable network passes in front of approximately 100 million homes, a respectable potential market for cable-based data services.

Cable modems provide an affordable option to achieve high-speed access to the Web, with current monthly subscription rates in the neighborhood of $40. They offer asymmetric access, that is, a much higher downstream speed than upstream, but for the majority of users this does not represent a problem since the bulk of their use will be for Web surfing during which the bulk of the traffic travels in the downstream direction.

When a customer uses a cable modem to access the Web, a splitter at the service demarc point directs the signal to the cable modem, usually located adjacent to the user's PC. The cable modem, in turn, connects to an Ethernet card that must be installed in the PC, and although all this can all be accomplished by a reasonably knowledgeable subscriber, the work is usually done by a cable installer (the proverbial cable guy) for roughly $100 in installation fees. In most areas where the service is available, the cable modem is provided as part of the service package. The 10 Base-T Ethernet card may or may not be included. If not, it can be had for approximately $30.

Although cable modems speed up access to the Web and other online services by hundreds of percentage points, a number of downsides must be considered. First and foremost, cable modem service is targeted at the consumer market, not at the business user. As a result, service quality and repair intervals are not necessarily what most corporate customers expect from a service that their business depends upon. Second, business customers who have a pre-established relationship with an ISP and an already well-known and advertised e-mail address will have to change them, because cable modem service is sold as part of an overall package, again targeted at the consumer market (in reality, the subscriber does not have to change; they would, however, have to pay for two accounts). This reality is currently under attack by such online services as *America Online* (AOL) who believe that as common carriers, the cable providers should be required to open their networks to all. The cable providers, in turn, argue that their own ISPs would be endangered if they were to allow such incursion. The courts are currently hearing arguments from both sides, and a decision will be forthcoming soon—stay tuned.

Another issue with cable modem service has to do with performance and is a function of the architecture of the cable system. In HFC architectures, the available bandwidth is shared among all subscribers attached to a particular node, hundreds in some cases. As a consequence, heavy usage by multiple subscribers could result in a serious slowdown in service, a phenomenon that is familiar to many cable subscribers today. Cable companies claim that they will overcome the problem through capacity buildouts. Time will tell.

The final concern that has been voiced about cable modems is one of security. Because the bandwidth is shared, it is conceivable that other subscribers could access the contents of other systems on that shared facility. And although this has not yet become a serious concern, it is valid. Cable modem subscribers should ensure that all utilities such as print and file sharing are disabled on their PCs to prevent this from happening.

One additional security issue that should be mentioned is that with a cable modem, the subscriber now has full-time access to the network and in many cases a fixed IP address. This potentially makes the PC as susceptible to attack by anyone on the network as any server.

DATA OVER CABLE STANDARDS

As interest grew in the late 1990s for broadband access to data services over cable television networks, CableLabs®, working closely with the ITU and major hardware vendors, crafted a standard known as the *Data Over Cable Service Interface Specification* (DOCSIS). Designed to ensure interoperability among cable modems as well as to assuage concerns about data security over shared cable systems, DOCSIS has done a great deal to resolve marketplace issues.

Under DOCSIS, CableLabs crafted a cable modem certification standard called DOCSIS 1.0 that guarantees that modems carrying the certification will interoperate with any headend equipment, are ready to be sold in the retail market, and interoperate with other certified cable modems. Engineers from Askey, Broadcom, Cisco Systems, Ericsson, General Instrument, Motorola, Philips, 3Com, Panasonic, Digital Furnace, Thomson, Terayon, Toshiba, and Com21 participated in the development effort.

The DOCSIS 1.1 specification was released in April 1999 and included two additional functional descriptions that began to be implemented in 2000. The first specification details procedures for guaranteed bandwidth as well as a specification for QoS guarantees. The second specification is called *Baseline*

Privacy Interface Plus (BPI+). It enhances the current security capability of the DOCSIS standards through the addition of digital certificate-based authentication and support for multi-cast services to customers. Although the DOCSIS name is in widespread use, CableLabs now refers to the overall effort as the CableLabs Certified Cable Modem Project.

DIGITAL SUBSCRIBER LINE (DSL)

The access technology that has enjoyed the greatest amount of attention recently is the *Digital Subscriber Line* (DSL). It provides an ideal solution for remote LAN access, Internet surfing, and access for telecommuters to corporate databases.

DSL came about largely as a direct result of the Internet's appearance in the public's consciousness in 1993. Prior to its arrival, the average telephone call lasted approximately four minutes, a number that central office personnel used while engineering switching systems to handle expected call volumes. This number was arrived at after nearly 125 years of experience designing networks for voice customers. They knew about Erlang theory, loading objectives, peak-calling days/weeks/seasons, and had decades of trended data to help them anticipate load problems. So, in 1993, when the Internet—and with it, the World Wide Web—arrived, network performance became unpredictable as callers began to surf, often for hours on end. The average four-minute call became a thing of the past as online service providers such as AOL began to offer flat rate plans that did not penalize customers for long connect times. Then the unthinkable happened: switches in major metropolitan areas, faced with unpredictably high call volumes and hold times, began to block[12] during normal business hours, a phenomenon that only occurred in the past during disasters or on Mother's Day. Something had to be done.

---

[12]Switches are designed in such a way that they will block the processing new incoming calls when they find themselves processing unusually high call volumes. This is done to ensure that emergency services can be maintained. Although this is a process that happens by design, under normal circumstances, it rarely occurs.

A number of solutions were considered, including charging different rates for data calls than for voice calls, but none of these proved feasible until a technological solution was proposed. That solution was DSL.

It is a well-known fact among telephony personnel that the analog telephone network has been carefully designed around the known behavior patterns of the human voice. As we noted earlier (again, thanks to work done at Bell Laboratories), the human voice comprises a rich mixture of frequencies that range from approximately 20 Hz to 20 KHz. And although this broad collection of spectral components adds tremendous richness, tone, and timbre to the sound of the voice, the extremely high and low frequency components are not necessary for the human ear to recognize and understand what is being said by another person. In fact, a relatively narrow range of frequencies is required. This range, known as the voice band, spans the spectrum of frequencies that fall between 300 and 3,300 Hz. As long as a telephone network can collect and deliver this portion of a caller's voice, the listener will be able to recognize and understand what the caller is saying.

This recognition was a real boon for the phone company, because it meant that they could engineer their networks with far better frequency efficiency. Instead of building their systems with the intent of giving every subscriber a 20-KHz piece of the spectrum, they were able to reduce that number to four KHz, thus allowing significantly more customers to be squeezed into whatever frequency range they had available. Thus, a 20-KHz piece of bandwidth could support approximately five customers instead of one.

It is commonly believed that because of the way the telephone network is designed, the local loop is incapable of carrying more than the frequencies required to support the voice band. This is patently untrue. ISDN, for example, requires significantly high bandwidth to support its digital makeup. When ISDN is deployed, the loop must occasionally be modified in certain ways to eliminate its designed bandwidth limitations. For example, some long loops are deployed with devices called *load coils* that "tune" the loop to the voice band (they are effectively notch filters). They make the transmission of frequencies above the voice band impossible but enable the relatively low-frequency voice band components to be carried across a long local loop. These load coils must be removed if digital services are to be deployed.

Thus, a local loop is only incapable of transporting high-frequency signal components if it is *designed* not to carry them. The capacity is still there; the network simply makes it unavailable. DSL services, especially ADSL, take advantage of this "disguised" bandwidth.

DSL TECHNOLOGY

While IDSL, SDSL, and HDSL are truly digital technologies, Asymmetric Digital Subscriber Line (ADSL), the best known form of DSL, is analog technology. The devices installed on each end of the circuit are sophisticated high-speed modems that rely on complex encoding schemes to achieve the high bit rates that DSL offers. Furthermore, several of the DSL services, specifically ADSL, g.lite, *Very High-Speed Digital Subscriber Lines* (VDSLs), and *Rate Adaptive Digital Subscriber Lines* (RADSLs) are designed to operate in conjunction with voice across the same local loop. ADSL is the most commonly deployed service and offers a great deal to both business and residence subscribers.

ADSL MODULATION TECHNIQUES

Two modulation techniques have emerged for the encoding of ADSL signals: *Discrete Multitone Modulation* (DMT) and *Carrier-Suppressed Amplitude Phase Modulation* (CAP). ISDN BRI's *2 Binary 1 Quaternary* (2B1Q) line code was an early entrant in the DSL world as the standard for *ISDN DSL* (IDSL), but it has become more of a niche scheme today compared to the other two.

CROSSTALK ISSUES

Because many DSL services use the standard local loop wire pair, there must be some technique in place to prevent crosstalk (channel bleeding) between the upstream and downstream sides of the circuit. Two techniques are commonly used: FDM, in which separate frequency ranges are established for the two transmitted components, and FDM combined with echo cancellation. In FDM systems, voice is relegated to a 0-to-4-KHz band, upstream traffic to a 25-to-200-KHz band, and downstream traffic to a 200-KHz-to-1.1-MHz band. Thus, the three components are completely isolated from one another. In echo-cancelled systems, the voice traffic remains in the 0-to-4-KHz band, but the upstream and downstream traffic share access to the 25-KHz-to-1.1-MHz band, hence the need for echo cancellation.

Echo cancellation is required whenever both ends of a circuit share access to the same range of available bandwidth. When one end transmits to the other end, it is common for some of the transmitted signal to be reflected back to the transmitter because of impedance mismatches. When this happens, the reflected signal can be mistaken for information received from the far end. To combat this, receivers remove the signal transmitted from anything they receive to ensure that what they receive is a pure transmitted signal from the far end. Generally speaking, CAP-based systems use FDM, while DMT-based systems combine FDM with echo cancellation.

DISCRETE MULTITONE MODULATION (DMT)

Discrete Multitone Modulation (DMT) is a relatively new encoding scheme that divides the available bandwidth into 256 4-KHz subcarriers. The information to be transmitted is then encoded into these subchannels based upon the relative performance of each channel. In other words, before information is encoded into the subcarriers, they are tested for relative performance based on noise and so on. Those with high noise coefficients are not utilized. Obviously, the more "clean" channels that are available, the higher the aggregate bandwidth. This results in an extremely granular assignment of bandwidth and high transmission quality.

DMT has been designated by the *American National Standards Institute* (ANSI) as the standard for ADSL (ANSI

T1.413-1995) and as such enjoys significant popularity throughout the industry as an open standard.

CARRIER-SUPPRESSED AMPLITUDE PHASE MODULATION (CAP)

Carrier-Suppressed Amplitude Phase Modulation (CAP) is a derivative of a well-known encoding scheme called *Quadrature Amplitude Modulation* (QAM, pronounced "kwam"). It uses a combination of amplitude and phase modulation to achieve high bits-per-signal encoding levels. CAP has the advantages of being more mature in the marketplace than DMT, significantly simpler than DMT, and better at echo cancellation, but it suffers from the serious disadvantage of being available from only a single source, GlobeSpan Semiconductor Corporation. Nevertheless, there is a significant amount of support for CAP in the marketplace in spite of the fact that it is a proprietary solution.

Enough about technology. On to services.

XDSL SERVICES

DSL comes in a variety of flavors, all designed to provide flexible, efficient, high-speed service across the existing telephony infrastructure. From a consumer point of view, DSL, especially ADSL, offers a remarkable leap forward in terms of available bandwidth for broadband access to the Web. As content has steadily moved away from being largely text-based and has become more image-based, the demand for faster delivery services is a cry that has been growing in intensity for some time now. DSL may provide the solution at a reasonable cost to both the service provider and the consumer.

Businesses, on the other hand, also stand to benefit from DSL technology rollouts. Remote workers and telecommuters can rely on DSL for remote LAN and Internet access, as well as for specialized services. A medical specialist can use DSL to view digitized medical images, a remote architect to view complex CAD files, or a film editor to review a digitized film clip. Furthermore, DSL provides a good technology solution for VPN access as well as for ISPs looking to increase the bandwidth available to their customers. DSL is available in a variety of both symmetric and asymmetric services, and thus offers a high-bandwidth access solution for a variety of applications. The most common DSL services are ADSL, *High Bit Rate Digital Subscriber Line* (HDSL), HDSL-2, RADSL, and VDSL. The special case of g.lite, a form of ADSL, will also be discussed.

ASYMMETRIC DIGITAL SUBSCRIBER LINE (ADSL)

When the Web and flat rate access charges arrived in all their splendor, the typical consumer phone call went from roughly four minutes in duration to several times that. All the engineering that had led to the overall design of the network based on an average four-minute hold time went out the window as the switches staggered under the added load. Never was the expression "In its success lie the seeds of its own destruction" more true. When ADSL arrived, it provided the offload that was required to save the network.

The typical ADSL installation is shown in Figure 1-12. No change is needed to the two-wire local loop, but minor equipment changes, however, are required. First, the customer must have an ADSL modem at their premises. This device enables both the telephone service and a data access device, such as a PC, to be connected to the line.

The ADSL modem is more than a simple modem in that it also provides the FDM process that is required to separate the voice and data traffic for transportation across the loop.[13] When voice traffic reaches the ADSL modem, it is immediately encoded in the traditional voice band frequencies that are

---

[13]The device that actually does this is called a *splitter*, in that it splits the voice traffic away from the data. It is usually bundled as part of the ADSL modem, although it can also be installed as a card in the PC, as a standalone device at the demarc point, or as a series of distributed devices on each phone at the premises. The most common implementation today is to have the splitter integrated as part of the DSL modem. This, however, is the least desirable implementation because this design can lead to crosstalk between the voice and data circuitry inside the device.

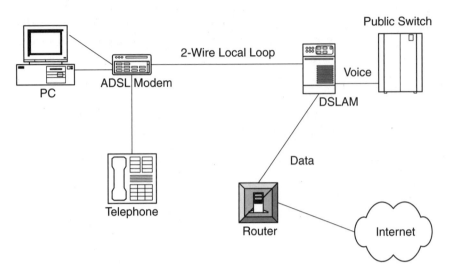

FIGURE 1-12 ADSL circuit layout

handed off to the local switch upon arrival at the central office. The modem is often referred to as an *ADSL Transmission Unit for Remote use* (ATU-R). Similarly, the device in the central office often called an ATU-C (for central office).

When a PC is used to transmit data across the local loop for Internet access or some other application, the traffic is encoded in the higher frequency band that is reserved for data traffic. The ADSL modem knows to do this because the traffic is arriving on a port that is reserved for data devices. Upon arrival at the central office, the data traffic does not travel to the local switch. Instead it stops at the ADSL modem that has been installed at the central office end of the circuit. In this case, the modem serves a large number of subscribers and is known as a *Digital Subscriber Line Access Multiplexer* (DSLAM, pronounced "dee-slam").

This is where the magic of DSL really shines. Instead of traveling on to the local switch, the data traffic is passed around the switch, a process known as a *line-side redirect*. The traffic is passed to a router, which in turn is connected to the Internet.

This architecture has a great deal of advantages. First, it offloads the data traffic from the local switch, which can then go back to doing what it does best, switching voice traffic. Second, it creates a new line of business for the service provider. As a result of adding the router and connecting the router to the Internet, the service provider instantly becomes an ISP. In keeping with the spirit of convergence, this is a near-ideal combination, because it enables the service provider to become a true service provider by offering much more than simple access and transport.

As the name implies, ADSL provides 2-wire asymmetric services; that is, the upstream bandwidth is different from the downstream. In the upstream direction, data rates vary from 16 to 640 Kbps, while the downstream bandwidth varies from 1.5 to 8 Mbps. Because most applications today are asymmetric in nature, this disparity poses no problem for the average consumer.

A Word About the DSLAM

This device has received a significant amount of attention recently because of the central role that it plays in the deployment of broadband access services. Obviously, the DSLAM must interface with the local switch so that it can pass voice calls on to the PSTN. However, it often interfaces with a number of other devices as well. For example, on the customer side, the DSLAM may connect to a standard ATU-C, directly to a PC with a built-in *Network Interface Card* (NIC), to a variety of DSL services, or to an integrated access device of some kind. On the trunk side (facing the switch), the DSLAM may connect to IP routers as described before, to an ATM switch, or to some other broadband service provider. It therefore becomes the focal point for the provisioning of a wide variety of access methods and service types.

High Bit Rate Digital Subscriber Line (HDSL)

The greatest promise of the *High Bit Rate Digital Subscriber Line* (HDSL) is that it provides a mechanism for the deployment of four-wire T1 and E1 circuits without the need for span repeaters, which can add significantly to the cost of deploying data services. It also means that service can be deployed in a matter of days rather than weeks, something customers certainly applaud.

DSL technologies in general enable repeaterless runs of wire as far as 12,000 feet, while traditional 4-wire data circuits such as T1 and E1 require repeaters every 6,000 feet. Consequently, many telephone companies are now deploying HDSL "behind the scenes" as a way of deploying these traditional services. Customers do not realize that the T1 facility they are plugging their equipment into is being delivered using HDSL technology. The important thing is that they don't *need* to know. All the customer should have to care about is that there is now a SmartJack installed in the basement, and through that jack they have access to 1.544 Mbps or 2.048 Mbps of bandwidth, period.

HDSL-2

HDSL-2 offers the same service that HDSL offers with one added (and significant) advantage. It does so over a single pair of wire, rather than two. It also provides other advantages. First, it was designed to improve vendor interoperability by requiring less equipment at either end of the span (such as transceivers and repeaters). Second, it was designed to work within the confines of standard telephone company *carrier serving area* (CSA) guidelines by offering a 12,000-foot wire-run capability that matches the requirements of CSA deployment strategies (see the discussion on CSA guidelines later in this section.)

A number of companies, including *Competitive Local Exchange Carrier* (CLEC) Rhythms NetConnections, have deployed T-1 access over HDSL-2 at rates that are 40 percent lower than typical T-Carrier prices. Allegiance Telecom is working with equipment manufacturer Adtran to determine the range of services that HDSL-2 will provide when deployed. Furthermore, a number of vendors including 3Com, Level One Communications, FlowPoint, Netopia,

Nortel Networks, and Alcatel have announced their intent to work together to achieve interoperability among DSL modems by working with the University of New Hampshire interoperability test labs.

RATE-ADAPTIVE DIGITAL SUBSCRIBER LINE (RADSL)

The *Rate-Adaptive Digital Subscriber Line* (RADSL, pronounced "Rad-zel") is a variation of ADSL designed to accommodate changing line conditions that can affect the overall performance of the circuit. Like ADSL, it relies on DMT encoding, which selectively "populates" subcarriers with transported data, thus allowing for granular rate-setting.

VERY HIGH-SPEED DIGITAL SUBSCRIBER LINE (VDSL)

The *Very High-Speed Digital Subscriber Line* (VDSL) is the newest DSL entrant in the bandwidth game and shows promise as a provider of extremely high levels of access bandwidth, as much as 52 Mbps over a short local loop. VDSL requires *Fiber-to-the-Curb* (FTTC) architecture and recommends ATM as a switching protocol. From a fiber hub, copper tail circuits deliver the signal to the business or residential premises. Bandwidth available through VDSL ranges from 1.5 to 6 Mbps on the upstream side, and from 13 to 52 Mbps on the downstream side. Obviously, the service is distance-sensitive and actual achievable bandwidth drops as a function of distance. Nevertheless, even a short loop is respectable when such high bandwidth levels can be achieved. With VDSL, 52 Mbps can be reached over a loop length of up to 1,000 feet, a not unreasonable distance by any means.

G.LITE

Because the installation of splitters has proven to be a contentious and problematic issue, the need has arisen for a version of ADSL that does not require splitters. This version is known as either ADSL Lite or g.lite (after the ITU-T G-Series standards[14] that govern much of the ADSL technology). In 1997, Microsoft, Compaq, and Intel created the *Universal ADSL Working Group* (UAWG),[15] an organization that grew to nearly 50 members dedicated to the task of simplifying the roll-out of ADSL. In effect, the organization had four stated goals:

· To ensure that analog telephone service will work over the g.lite deployment without remote splitters, in spite of the fact that the quality of the voice may suffer slightly due to the potential for impedance mismatch

· To maximize the length of deployed local loops by limiting the maximum bandwidth provided. Research indicates that customers are far more likely to notice a performance improvement when migrating from 64 Kbps to 1.5 Mbps than when going from 1.5 Mbps to higher speeds. Perception is clearly important in the marketplace, so the UAWG chose 1.5 Mbps as their downstream speed.

· To simplify the installation and use of ADSL technology by making the process as "plug-and-play" as possible

· To reduce the cost of the service to a perceived reasonable level

Most DSL vendors are working on ways to achieve success with DSL in the marketplace, and g.lite is high on various lists as a winning option. At the 1999 annual DSL Conference and Exhibition in Reston, Virginia, Steve Rago of Lucent Technologies identified six key challenges that must be overcome if DSL is to be a viable contender for broadband access. They are

---

[14]ITU-T stands for *International Telecommunications Union-Telecommunications Standardization Sector* and is the organization that replaced the *International Telegraph and Telephone Consultative Committee* (CCITT) in 1992.

[15]The group self-dissolved in the summer of 1999 after completing what they believed their charter to be.

- Service provisioning
- Reliability
- Suitability of loop plant
- Subscriber scalability
- Power management
- Performance

Ultimately, we will see the first two challenges addressed through the use of integrated voice and data line cards. The second two will be overcome through the deployment of both g.lite and full-service ADSL, depending on the nature of the loop. The final two will be achieved through the deployment of superior microelectronics.

Of course, g.lite is not without its detractors. A number of vendors have pointed out that if g.lite requires the installation of microfilters at the premises on a regular basis, then true splitterless DSL is a myth, since microfilters are in effect a form of splitter. They contend that if the filters are required anyway, then they might as well be used in full-service ADSL deployments to guarantee high-quality service delivery. Unfortunately, this flies in the face of one of the key tenets of g.lite, which is to simplify and reduce the cost of DSL deployment by eliminating the need for an installation dispatch (a "truck roll" in the industry's parlance). The key to g.lite's success in the eyes of the implementers is to eliminate the dispatch, minimize the impact on traditional *plain old telephone service* (POTS) telephones, reduce costs, and extend the achievable drop length. Unfortunately, customers still have to be burdened with the installation of microfilters, and coupled noise on POTS is higher than expected. Many vendors argue that these problems largely disappear with full-feature ADSL using splitters. A truck dispatch is still required, but again, it is often required to install the microfilters anyway, so there is no net loss. Furthermore, a number of major semiconductor manufacturers support both g.lite and ADSL on the same chipset, so the decision to migrate from one to the other is a simple one that does not necessarily involve a major replacement of internal electronics.

DSL MARKET ISSUES

DSL technology offers advantages to both the service provider and the customer. The service provider benefits from successful DSL deployment because it serves not only as a cost-effective technique for satisfying the bandwidth demands of customers in a timely fashion, but also because it provides a Trojan horse approach to the delivery of certain pre-existing services. For example, many providers today implement T-1 and E-1 services over HDSL because it is cost-effective. And although HDSL-2 is not yet widely deployed, it soon will be, at which time the advantages for the service provider will be even more enhanced as a result of the two-wire, rather than four-wire, nature of the service. Customers are blissfully unaware of this. In this case, it is the service provider rather than the customer who benefits most from the deployment of the technology. Of course, the accelerated installation interval is very much to the customer's advantage.

From a customer point of view, DSL provides a cost-effective way to buy medium-to-high levels of bandwidth, and in some cases embedded access to content. In 1999, AOL announced strategic partnerships with both Bell Atlantic and SBC Corporation under which customers could buy DSL access to AOL, bundled for $40 per month. This proved to be a boon for both Bell Atlantic and SBC because it gave them the coveted relationship with a content provider that they desired. It also gave them a ready market for their DSL services, a market that they had been somewhat reluctant to enter because of the current regulatory environment. Under the terms of the tariffs that guide their activities in their operating regions, the *Incumbent Local Exchange Carriers* (ILECs, or the former RBOCs) must sell services like DSL at a discount to their wholesale customers, many of whom are their direct in-region competitors. The danger lies in the fact that those customer/competitors could easily buy unbundled DSL facilities from the incumbent company, form a long-distance relationship with a bandwidth provider like Qwest, and take the ILEC completely out of the service equation. This is clearly an undesirable outcome for the ILECs, so the opportunity to partner with the likes of AOL offers a somewhat protected point of penetration for DSL services.

Furthermore, the ILECs enjoy a substantial revenue stream from T-1 services that operate at 1.544 Mbps. If DSL services can provide the same bandwidth at a substantially lower cost to the customer, then T-1 revenues could be endangered by the widespread deployment of DSL. This concern is clearly in the forefront of the minds of the local incumbent service providers, but they also recognize that in this chaotic market it is sometimes necessary to cannibalize some traditional market segments in order to maximize long-term profit in others.

Other forces have affected the DSL marketplace as well. In November 1999, the FCC ruled that local telephone companies must unbundle their local loops to the extent that competitors can share access to the copper in order to provide DSL services, even if the local carrier provides the voice service over the same line. In other words, under the terms of the ruling, a customer in Dallas, for example, could buy their voice service from SBC and high-speed data access service (DSL) from Covad Communications Group, both delivered across the customer's existing local loop. The Commission's intent with the line sharing ruling is to increase the number of companies offering high-speed data services while at the same time accelerating and geographically broadening the deployment of DSL. ILECs already implement line sharing with their own voice and DSL offerings, and because DSL is a form of FDM, it lends itself to this kind of sharing arrangement among competitors that share access to a common physical infrastructure. Customers benefit from this arrangement because they are not required to order a second line for DSL if they choose to purchase the service from a competitive provider, whereas they were prior to the ruling. ILECs also benefit, because their own DSL services are still in the running as a competitive offering via the local loop.

DSL also represents the primary bastion of defense against the ongoing market penetration efforts of cable modems. Currently, cable modems are taking the broadband access market by storm because the technology is widely available and works reasonably well, noticeably better (read faster) than traditional analog modem technology. DSL has yet to achieve the penetration levels that cable modems enjoy. A number of technical issues including loop length and age of the outside plant are having a significant negative impact on widespread deployment. Cable modems are expected to lead the broadband access charge for the next few years, but DSL is expected to catch up and eventually bypass the level of cable penetration. In fact, the ILECs have for the most part made aggressive plans to roll out DSL on a large scale.

Another contributing factor in favor of the ILECs is ubiquity. The cable industry's network passes within service distance of more than 95 percent of the homes and businesses in the country. However, their actual market penetration is substantially lower than that. The telephone company's twisted pair network, on the other hand, is close to universally deployed, with substantially higher customer penetration levels. Furthermore, many ILECs have crafted service alliances with direct broadcast television satellite providers, and between the soon-to-be-available two-way data transmission that these satellites will provide and the in-place, high-speed service that DSL makes possible, the incumbents will most likely remain the provider of choice for both voice and data services. According to a recent Price-Waterhouse study on who customers would select as their residential service provider, the ILECs win the game:

ILEC	$\rightarrow$	40%
IXC	$\rightarrow$	21%
Satellite	$\rightarrow$	7%
CATV	$\rightarrow$	7%
Power	$\rightarrow$	5%
Wireless	$\rightarrow$	2%

Business customers, on the other hand, have expressed very different numbers in the study. Thirty-six percent of them indicate that they would switch to an *Interexchange Carrier* (IXC) or *Competitive Local Exchange Character* (CLEC) if given the opportunity, based purely on service, not technology. This is a challenge that the ILECs must remain cognizant of as they go forward with their competitive positioning plans.

PROVIDER CHALLENGES

The greatest challenges facing those companies looking to deploy DSL are competition from cable, unfettered pent-up customer demand, installation issues, and plant quality.

Competition from cable companies represents a threat to wireline service providers on several fronts. First, cable modems enjoy a significant amount of press and are therefore gaining well-deserved marketshare. The service they provide, although not always available, is well-received in the areas served and offers high-quality, high-speed access. Although cable modem service availability is considered to be spotty, so is DSL. According to estimates by IDC, cable modems owned 50 percent of the high-speed subscribers in 1998 compared to DSL's meager 5 percent. Although those numbers are expected to shift, DSL is expected to grow to 31 percent by 2002, while cable modems will achieve 59 percent penetration within the same timeframe. The law of primacy is well-demonstrated here. The first to market wins. Nevertheless, the incumbent providers may pull off a surprise attack by forming service alliances to garner a larger percentage of the market—we'll see.

The second challenge is unmet customer demand. If DSL is to satisfy the broadband access requirements of the marketplace, it must be made available throughout ILEC service areas. This means that incumbent providers must equip their central offices with DSLAMs that provide the line-side redirect required as stage one of DSL deployment. Again, the law of primacy rears its head here. The ILECs must get to market quickly with their own broadband offerings if they are to attain and retain a toehold in the burgeoning broadband access marketplace.

One advantage that will help them achieve primacy is the strategic relationship some ILECs have made with service providers. Bell Atlantic, GTE, and SBC have forged alliances with AOL, mentioned earlier, to provide DSL-based access to the service, bundled for about $40 per month. This form of technology and service convergence represents an advantage for everyone. Customers benefit from having one-stop shopping for both access and service, AOL gets enhanced access to ILEC customers, and the ILECs find a safe harbor within which to deploy DSL. Furthermore, AOL is aligned with DirecTV, DirecPC, and TiVo, all of which provide alternative wireless access methods. If the ILECs become part of that alliance, particularly in the face of cable's refusal to allow the AOLs of the world ready access to their networks, the ILECs and their partners could achieve marketplace primacy over the broadband cable providers. Again, time will tell. Manufacturers like Lucent have stepped up to the challenge of rapid deployment with such products as their Stinger™ Access Concentrator, a DSLAM-like, central office-based device that supports SDSL, ADSL, g.lite, and HDSL-2 . It offers a high-speed ATM switching fabric that enables service providers to offer diverse and granular levels of QoS, something the market is keen about. Stinger also offers support for both frame relay and ATM access and incorporates a "Wizard-based" configuration and network management system.

The third and fourth challenges to rapid and ubiquitous DSL deployment are installation issues and plant quality. A significant number of impairments have proven to be rather vexing for would-be deployers of widespread DSL. These challenges fall into two categories: electrical disturbances, and physical impairments.

ELECTRICAL DISTURBANCES

The primary effect of electrical disturbance in DSL is crosstalk, caused when the electrical energy carried on one pair of wires bleeds over to another pair and causes interference (noise). Crosstalk exists in several flavors. *Near-end crosstalk* (NEXT) occurs when the transmitter at one end of the link interferes with the signal received by the receiver at the same end of the link, while *far-end crosstalk* (FEXT) occurs when the transmitter at one end of the circuit causes problems for the signal received by a receiver at the far end of the circuit. Similarly, problems can occur when multiple DSL services of the same type exist in the same cable and interfere with one another. This is referred to as self-NEXT or self-FEXT. When different flavors of DSL interfere with one another, the phenomenon is called foreign-NEXT or foreign-FEXT. Other problems that can cause errors in DSL and therefore a limitation in the maximum achievable bandwidth of the system include simple *radio frequency interferences* (RFI) and impulse, Gaussian, and random noises that exist in the background but can affect signal quality even at extremely low levels.

PHYSICAL IMPAIRMENTS

The physical impairments that can have an impact on the performance of a newly-deployed DSL circuit tend to be characteristics of the voice network that typically have a minimal effect on simple voice and low-speed data services. These include load coils, bridged taps, splices, mixed gauge loops, and weather conditions.

LOAD COILS.   Load coils present a problem for DSL deployment because of their internal anatomy. Load coils are nothing more than "lumped inductance" installed on long loops to overcome distance-related impairments to voice. In effect, load coils are low-pass filters that tune the local loop to the voice band frequency range. Because the high-frequency components of the human voice are not required, the load coils filter them out, leaving nothing but the lower frequency voice band. Also, because low-frequency signals propagate farther than high-frequency signals,[16] the addition of load coils enables the frequencies required to understand and recognize the received

---

[16]Think about the last time that you attended a parade. You may recall that the event that notified you of the imminent arrival of the parade, led by the band, was the low-frequency sound of the bass drums, not the brass section. The drum sounds reached your ears first.

voice to be carried the maximum distance possible without worrying about the "smearing" that would occur if the higher frequency components were included in the mix.

Unfortunately, load coils effectively eliminate the high-frequency transmission channel required to transport the broadband data segment of the DSL service. They must therefore either be removed prior to deploying DSL across a loaded loop, or another loop that is not loaded must be used.

BRIDGED TAPS.   When a multi-pair telephone cable is installed in a newly built-up area, it is generally some time before the assignment of each pair to a home or business is actually made. To simplify the process of installation when the time comes, the cable's pairs are periodically terminated in terminal boxes (sometimes called *B-boxes*) installed every block or so. There may be multiple appearances of each pair on a city street, waiting for assignment to a customer. When the time comes to assign a particular pair, the installation technician will simply go to the terminal box where that customer's drop appears and cross-connect the loop to the appearance of the cable pair (a set of lugs) to which that customer has been assigned. This eliminates the need for the time-consuming process of splicing the customer directly into the actual cable.

Unfortunately, this efficiency process also creates a problem. Although the customer's service has been installed in record time, now unterminated appearances of the customer's cable pair exist in each of the terminal boxes along the street. These so-called bridged taps present no problem for analog voice, but they can be a catastrophic source of noise due to signal reflections that occur at the copper-air interface of each bridged tap. If DSL is to be deployed over the loop, the bridged taps must be removed. Although the specifications indicate that bridged taps do not cause problems for DSL, actual deployment says otherwise. To achieve the bandwidth that DSL promises, the taps must be terminated.

SPLICES AND GAUGE CHANGES.   Although they cause less of a problem than some of the other impairments, splices can result in service impairments due to cold solder joints, corrosion, or weakening effects caused by the repeated bending of wind-driven aerial cable.

Gauge changes tend to have the same problems that plague circuits with unterminated bridged taps. When a signal traveling down a wire of one gauge jumps to a piece of wire of another gauge, the signal is reflected, resulting in an impairment known as *intersymbol interference*. The use of multiple gauges is common in a loop deployment strategy because it enables outside plant engineers to use lower-cost, small gauge wire where appropriate, cross-connecting it to larger-gauge, lower-resistance wire where necessary.

WEATHER CONDITIONS.   Weather conditions is perhaps a bit of a misnomer. The real issue is moisture. One of the greatest challenges facing the deployment of DSL is the age of the loop plant. Much of the older distribution cable uses inadequate insulation between the conductors (paper in some cases). In certain cases, the outer sheath has cracked, allowing moisture to seep into the cable itself. The moisture causes crosstalk between wire pairs that can last until the water evaporates. Unfortunately, this can take a considerable amount of time and result in extended outages.

SOLUTIONS

All of these factors have solutions. The real question is whether they can be remedied at a cost that is within reasonably achievable bounds. Given the growing demand for broadband access, there seems to be little doubt that the elimination of these factors would be worthwhile at all reasonable costs, particularly considering how competitive the market for the local loop customer has become.

The electrical effects, largely caused by various forms of crosstalk, can be reduced or eliminated in a variety of ways. Physical cable deployment standards are already in place that, when followed, help to control the amount of near and far-end crosstalk that can occur within binder groups in a given cable. Furthermore, filters have been designed to eliminate the background noise that can creep into a DSL circuit.

Physical impairments can be controlled to a point, although, to a certain extent, the service provider is at the mercy of their installed network plant. Obviously, older cable will present more physical impairments than newer cable, but the service provider can take steps to maximize its success rate when entering the DSL marketplace. The first step is to prequalify local loops for DSL service to the greatest extent possible. This means running a series of tests using *mechanized loop testing* (MLT) to determine whether each loop's transmission performance falls within the bounds established by the service provider, existing standards, and industry support organizations.

For DSL prequalification, MLT tests the following performance indicators (and others as required):

· Cable architecture
· Loop length
· Crosstalk and background noise

PREVENTATIVE NETWORK ARCHITECTURE ISSUES

New cable architecture is related to the actual deployment plan of the loop being tested. Until the mid-1980s, a criterion known as the "Resistance Design Rule" dictated that the maximum loop resistance allowed would be 1,500 ohms, with no limit on the number of bridged taps or gauge changes. This made sense, considering that the plant was intended and designed for the transport of voice and low-speed, voice-band data.

After the mid-1980s, things changed. Instead of focusing exclusively on resistance as the primary loop design criterion, outside plant engineers shifted to the use of CSA guidelines, which limit both the number of gauge changes and the maximum loop length, as well as the maximum length of any bridged taps that are present. This design came in concert with the widespread deployment of digital loop carrier architectures, which came about as a way to reduce the cost of deploying service to far-flung parts of a service provider's coverage area. In CSA architectures, customers receive their service as one of a number of customers within a geographical area that is served by a remote multiplexer, known as a *remote terminal* (RT). All customers on an RT share access to the central office over a collection of shared wire pairs using traditional T-Carrier-like TDM.

The layout of a CSA is shown in Figure 1-13. The *central office terminal* (COT) is connected to the RT using a collection of standard 24-channel T-Carrier facilities. Because the T-Carrier is digital, it can be as long as required, using in-span repeaters to maintain signal quality. Thus, a CSA architecture can provide service to remote areas using significantly fewer wire pairs than would be required if each customer were given one. From the remote terminal, the local loops can run an additional 9,000 feet over 26-gauge wire or 12,000 feet over 24-gauge wire. Neither requires load coils over that distance.

From a DSL point of view, traditional CSA presents a problem. A T-Carrier allocates a single DS-0 channel to each customer, effectively limiting the maximum bandwidth achievable

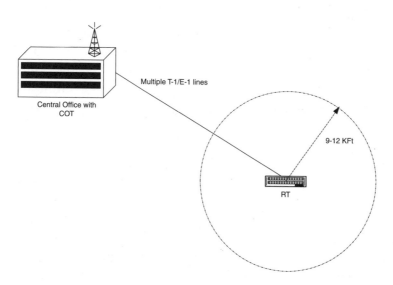

FIGURE 1-13 A digital loop carrier system

over that channel to 64 Kbps, far less than DSL requires. The good news is that so-called *Next Generation Digital Loop Carriers* (NGDLCs) stand to overcome this challenge. NGDLCs rely largely on a fiber-based infrastructure, which overcomes the bulk of the available bandwidth problem. The channelization issue is remedied by the deployment of Bellcore Standard GR-303, which gives NGDLCs the capability to dynamically assign available bandwidth on an as-needed basis. Bellcore estimates that nearly half of all local loops will be served by DLCs this year, and the number will grow. Given GR-303's capability to provision bandwidth in increments greater than 64 Kbps and DSL's demand for high bandwidth, CSA architectures designed around NGDLC technology are well positioned to help bring about the widespread deployment of DSL.

LOOP LENGTH

DSL chipsets tend to be extremely sensitive to loop length. Most of the symmetric DSL services operate at distances of 12,000 feet, while the asymmetric services operate up to 18,000 feet across the loop. As long as deployment is limited to a local loop plant that is within the distance range stipulated by the DSL standards, the service stands a good chance of working. However, many of the existing chipsets are so distance-sensitive that they will fail to work properly over loops that are as little as 650 feet longer than the stipulated maximum length. Obviously, there is nothing that can be done to physically shorten a local loop, but NGDLC technology may provide the solution.

CROSSTALK AND BACKGROUND NOISE

Crosstalk can be minimized by ensuring to the greatest degree possible that only one type of DSL be deployed within a binder group[15] to minimize the amount of foreign near-end crosstalk. Background noise is another matter entirely that, to a large extent, is unpredictable and therefore difficult to manage. Some electrical noise such as that caused by electric motors, air conditioner and refrigerator compressors, and other RFI sources can be managed through the judicious use of noise filters. Thermal noise, on the other hand, is caused by variations in ambient temperature, and short of refrigerating the cables, there is little that can be done to prevent it. It must therefore largely be worked around rather than eliminated.

DOWN THE ROAD

A significant amount of energy is being funneled into DSL services and technology development. *Integrated Telecom Express* (ITEX) of Silicon Valley has announced a chipset that will enable DSL services to be deployed across loops as long as 20,000 feet, far longer than the DSL standards stipulate. This will enable service providers to reach customers who are currently on the fringe of their service area for DSL. This is clearly a development to watch.

The integrated access devices market is also afire. In response to customer demand from smaller companies for the ability to move integrated voice and data over a single facility, a number of vendors including Sylantro, CopperCom, TollBridge Technologies, and Accelerated Networks have announced voice-over DSL solutions. FlowPoint, a subsidiary of Cabletron, along with Lucent and Copper Mountain, has announced their integrated access devices that deliver multiple lines and services across a single local loop. Using ATM as the transport mechanism, the so-called *Next Generation Integrated Access Devices* encapsulate IP-based data into ATM cells and provide the DSL interface for voice traffic. The *Integrated Access Device* (IAD)

---

[17]A binder group is a group of 25 cable pairs within a larger cable.

manages all functions previously handled by a DSL modem as well as a bridge or router, and because it relies on ATM transports, it can take advantage of the granular QoS control that ATM enables. Once the traffic reaches the service provider's central office, a DSLAM routes data traffic to the data network and encapsulated voice traffic to a voice gateway where it is converted to a standard voice signal and handed off to a local switch.

The Data CLECs certainly believe in the capabilities of DSL technology, and the market seems to believe in *them*. Three of them, Rhythms NetConnections, Covad Communications, and NorthPoint, succeeded with IPOs and as of August 1999 were valued in the $4 to 5 billion range on 1999 sales projections of $10 to $50 million. To add to that success, major ISPs have lined up to partner with the data CLECs to be able to offer higher-speed access to their business customers. Furthermore, major bandwidth players in the industry such as Qwest have stepped up to form long-term relationships with these companies, and some have gone so far as to invest in them. In January 1998, AT&T invested $25 million in Covad Communications, and MCI WorldCom invested $30 million in Rhythms NetConnections.

DSL SUMMARY

From a service convergence perspective, DSL represents a major component in the evolving full-service network. Customer requirements are best met when the solution offered to them is a simple one that integrates all of their needs on a common infrastructure or platform. If service providers can overcome the technical issues that currently hinder widespread deployment of DSL (and they certainly will), it will become a powerful force in the modern network and will ascend to the point of primacy in the battle for broadband access control.

Furthermore, there is a management issue that must be addressed by would-be providers. It is far better to deny DSL service to a customer initially than to sell the service and later discover that it cannot be turned up on the customer's line because of inadequate loop performance. Service providers have therefore created processes to minimize the inconvenience to the customer by following a two-step approach. First, they test local loops that fall within the service area for potential DSL deployment using MLT and manual testing procedures to create databases that can be quickly consulted when a request for service comes in. Second, if the loop does qualify for DSL, they test the service carefully after turnup to ensure that the loop performs properly across the entire range of frequencies required by both the voice and data signal components. This process first ensures that the service is deliverable and secondly guarantees that it works when turned over to the customer, all without burdening the customer unnecessarily with technology barriers that they have no reason to be concerned with.

WIRELESS ACCESS TECHNOLOGIES

It is only in the last few years that wireless access technologies have advanced to the point that they are being taken seriously as contenders for the broadband local loop market. Traditionally, minimal infrastructures were in place, and they were bandwidth-bound and error-prone to the point that wireless solutions were not considered as serious contenders.

Wireless access technologies have undergone an evolution comprising three generations. First-generation systems, which originated in the late 1970s and throughout the 1980s, were entirely analog in nature and supported almost exclusively voice with very little data. They are characterized by the use of *Frequency Division Multiple Access* (FDMA) technology. In FDMA, users are assigned analog frequency pairs over which they send and receive. One frequency serves as a voice transmit channel, the other as a receive channel.

Second-generation systems, which came about in the 1990s, were all digital and were still primarily voice-oriented, although data transport became more accepted. In second-generation systems, digital access became the norm through such technologies as *Time Division Multiple Access* (TDMA), *Global System for Mobile Communications* (GSM), and *Code*

Division Multiple Access (CDMA). In TDMA systems, the available frequency is broken into channels as in FDMA, but here the similarity ends. The channels are shared among a group of users who use a time division technique to ensure fair and equal access to the available channel pairs.

In CDMA, sometimes called *Spread Spectrum*, there is no channelization per se. Instead, all users share access to the entire range of the available spectrum, and special techniques are utilized to isolate one conversation from another. These techniques include frequency hopping, in which each conversation "hops" randomly from one frequency to another, and noise modulation, in which the signal is modulated against a pseudo-random noise signal so that it appears to be background noise to anyone listening in.

Finally, in *third-generation* (3G) systems, we see the emergence of true broadband access and the promise of high-speed data in addition to voice. Access technologies for broadband will include *Wideband CDMA* (W-CDMA) or an enhanced version of GSM. In late 1999, the ITU created a comprehensive set of third-generation standards designed to accommodate the various technological directions that implementers have taken and to ensure that current systems can gracefully evolve to new 3G standards.

Five terrestrial radio interface standards were accepted: IMT DS, also known as Wideband CDMA; IMT MC, also known as cdma2000; IMT TC, also known as UTRA TDD; IMT SC, also known as UWC-136 or EDGE; and finally, IMT FT, commonly known as DECT.

Recently, a new family of broadband wireless technologies has emerged that poses a serious threat to traditional wired access infrastructures. These include *Local Multipoint Distribution Service* (LMDS), *Multichannel Multipoint Distribution Service* (MMDS), and *Geosynchronous* (GEO) and *Low Earth Orbit* (LEO) satellites.

LOCAL MULTIPOINT DISTRIBUTION SERVICE (LMDS)

LMDS is a bottleneck resolution technology designed to alleviate the transmission restriction that occurs between high-speed

LANs and WANs. Today local networks routinely operate at speeds of 100 Mbps (Fast Ethernet) and even 1,000 Mbps (Gigabit Ethernet), which means that any local loop solution that operates slower than either of those poses a restrictive barrier to the overall performance of the system. LMDS offers a good alternative to wired options. Originally offered as CellularVision, it was seen by its inventor, Bernard Bossard, as a way to provide cellular television as an alternative to cable. In August of 1993, Bell Atlantic purchased an interest in the company and currently runs a multi-cell system in Brooklyn.

Operating in the 28-GHz range, LMDS offers data rates as high as 155 Mbps, the equivalent of SONET OC-3c. Because it is a wireless solution, it requires a minimal infrastructure and can be deployed quickly and cost-effectively as an alternative to the wired infrastructure provided by incumbent service providers. After all, the highest cost component when building networks is not the distribution facility, but rather the labor required to trench it into the ground or build aerial facilities. Thus, any access alternative that minimizes the cost of labor will garner significant attention.

LMDS relies on a cellular-like deployment strategy under which the cells are approximately three miles in diameter. Unlike cellular service, however, users are stationary. Consequently, there is no need for LMDS cells to support roaming. Antenna/transceiver units are generally placed on rooftops, as they need an unobstructed line of sight to operate properly. In fact, this is one of the disadvantages of LMDS (and a number of other wireless technologies). Besides suffering from unexpected physical obstructions, the service also suffers from "rain fade" caused by the absorption and scattering of the transmitted microwave signal by atmospheric moisture. Even some forms of foliage will cause interference for LMDS, so the transmission and reception equipment must be mounted high enough to avoid such obstacles, hence the tendency to mount the equipment on rooftops.

Because of its high bandwidth capability, many LMDS implementations interface directly with an ATM backbone to take advantage of both its bandwidth and its diverse QoS capabilities. If ATM is indeed the transport fabric of choice, then the

LMDS service becomes a broadband access alternative to a network capable of transporting a full range of services including voice, video, image, and data—the full suite of multimedia applications.

THE LMDS MARKET

In addition to Bell Atlantic, which bought an interest in CellularVision early on, other players have also stepped into the fray. Lucent, Broadband Networks, Texas Instruments, Bosch, Cisco, and Hewlett-Packard have all committed themselves to the game, and a number of service providers have adopted LMDS as their broadband wireless local loop solution. In July 1998, CellularVision agreed to sell 850 of its 1,300 LMDS licenses in New York City for $32.5 million to Winstar Communications, which will use them to deliver local telecommunications services within the greater New York City area. Although this may seem like a rather extreme step, consider the fact that as many as 65 percent of businesses in major cities are not served by fiber. This makes LMDS an ideal access scheme for broadband delivery.

Three key applications for LMDS have emerged in markets where the technology has been deployed: voice, remote LAN access and interconnection, and interactive television. These three applications alone make LMDS a powerful contender in the broadband access market.

MULTICHANNEL, MULTIPOINT DISTRIBUTION SYSTEM (MMDS)

MMDS got its start as a "wireless cable television" solution. In 1963, a spectrum allocation known as the *Instructional Television Fixed Service* (ITFS) was carried out by the FCC as a way to distribute educational content to schools and universities. In the 1970s, the FCC established a two-channel metropolitan distribution service called the *Multipoint Distribution Service* (MDS). It was to be used for the delivery of pay-TV signals, but with the advent of inexpensive satellite access and the ubiquitous deployment of cable systems, the need for MDS disappeared.

In 1983, the FCC rearranged the MDS and ITFS spectrum allocation, creating 20 ITFS education channels and 13 MDS channels. In order to qualify to use the ITFS channels, schools had to use a minimum of 20 hours of airtime, which meant that ITFS channels tended to be heavily, albeit randomly, utilized. As a result, MMDS providers that use all 33 MDS and ITFS channels must be able to dynamically map requests for service to available channels in a completely transparent fashion, which means that the bandwidth management system must be reasonably sophisticated.

Because MMDS is not a true cable system (in spite of the fact that it has its roots in television distribution), no franchise issues exist for its use (of course, licensing requirements do exist). However, the technology is also limited in terms of what it can do. Unlike LMDS, MMDS is designed as a one-way broadcast technology and therefore does not typically enable upstream communication. Beginning in November 1997, Bellsouth deployed an MMDS-based home entertainment system in New Orleans, offering more than 160 channels of Americast programming. MMDS enables transmission distances of up to 35 miles, much farther than LMDS, and by December 1998, the company had a similar presence in Atlanta, Charleston, Birmingham, and Jacksonville as well. Today their coverage area has expanded even more dramatically, and other companies have also entered the game.

However, many contend that there is adequate bandwidth in MMDS to provision two-way systems, which would make it suitable for voice, Internet access, and other data-oriented services. In fact, a number of providers have crafted two-way systems that are performing rather well. American Telecasting, an MMDS-based wireless cable television company, enjoys a great deal of success in their market, so much in fact that in April 1999 Sprint announced plans to acquire the company and use its spectrum allocation to provide wireless access to the Sprint ION network for high-speed Internet and other data applications.

Their acquisitions did not stop there, however. In addition to American Telecasting, Sprint owns Videotron USA and WBS America. Sprint will use the technology in such major market areas as Denver, Portland, Las Vegas, and Seattle.

Similarly, in May 1999, CAI Wireless, a wireless cable provider and the owner of MMDS-dependent CS Wireless Systems, announced their intent to be acquired by MCI Worldcom for approximately $475 million. CS Wireless became one of the largest users of MMDS in the world, operating in 11 markets. Unfortunately, wireless cable services provided by New York-based CAI Wireless and Plano-based CS Wireless did not prove to be as profitable as had been expected, and the company filed for bankruptcy in 1998. It is expected by many that MCI Worldcom will use the MMDS spectrum for broadband wireless Internet access, similar to Sprint's plans for American Telecasting's holdings.

Clearly, companies like Sprint and MCI Worldcom see the value of wireless access to their own networks and have taken steps to bring about the technological convergence of innovative technologies to meet the growing customer demand for high-speed Internet access. The manufacturing sector certainly sees advantages in the LMDS and MMDS markets. More than a dozen companies now manufacture equipment for the two technologies including such heavyweights as Lucent Technologies, Nortel Networks, and Newbridge Networks, to name a few. Furthermore, both Cisco and Motorola have thrown their hats in the ring. In June 1999, the two companies announced their joint intent to acquire Bosch Telecom's LMDS division.

SO WHAT'S THE MARKET?

Service providers looking to sell LMDS and MMDS technologies into the marketplace should target small and medium-size businesses that experience measurable peak data rates and are looking to move into the packet-based transport arena. Typical applications for the technologies include LAN interconnections, Internet access, and cellular backhauls between *mobile telephone switching offices* (MTSOs) and newly deployed cell sites. LMDS tends to offer higher data rates than MMDS; LMDS peaks at a whopping 1.5 Gbps, while MMDS can achieve a maximum transmission speed of about three Mbps. Nevertheless, both have their place in the technology pantheon.

SATELLITE TRANSMISSION

> It will be observed that one orbit, with a radius of 42,000 km, has a period of exactly 24 hours. A body in such an orbit, if its plane coincided with that of the earth's equator, would revolve with the earth and would thus be stationary above the same spot on the planet. It would remain fixed in the sky of a whole hemisphere and unlike all other heavenly bodies would neither rise nor set. A body in a smaller orbit would revolve more quickly than the earth and so would rise in the west, as indeed happens with the inner moon of Mars . . .
>
> . . . Let us now suppose that such a station were built in this orbit. It could be provided with receiving and transmitting equipment (the problem of power will be discussed later) and could act as a repeater to relay transmissions between any two points on the hemisphere beneath, using any frequency which will penetrate the ionosphere. If directive arrays were used, the power requirements would be very small, as a direct line of sight transmission would be used. There is the further important point that arrays on the earth, once set up, could remain fixed indefinitely. Moreover, a transmission received from any point on the hemisphere could be broadcast to the whole of the visible face of the globe, and thus the requirements of all possible services would be met. (See Figure 1-14.)

In October 1945, Arthur C. Clarke published a paper in *Wireless World* entitled, "Extra-Terrestrial-Relays: Can Rocket Stations Give World-Wide Radio Coverage?" In his paper,

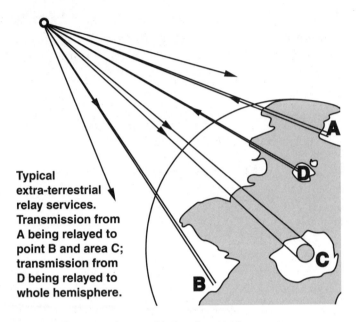

Typical
extra-terrestrial
relay services.
Transmission from
A being relayed to
point B and area C;
transmission from
D being relayed to
whole hemisphere.

FIGURE 1-14   Excerpted from *Wireless World*, October 1945

Clarke proposed the concept of an orbiting platform that would serve as a relay facility for radio signals sent to it that could be turned around and retransmitted back to Earth with far greater coverage than was achievable through terrestrial transmission techniques. His platform, described in the excerpt shown previously, would orbit at an altitude of 42,000 kilometers (25,200 miles) above the equator where it would orbit at a speed identical to the rotation speed of the earth. As a consequence, the satellite would appear to be stationary to Earth-bound users.

Satellite technology may prove to be the primary communications gateway for regions of the world that do not yet have a terrestrial wired infrastructure, particularly given the fact that they are now capable of delivering broadband services. In addition to the U.S., the largest markets for satellite coverage are Latin America and Asia, particularly Brazil and China. According to the Strategis Group's second edition of *World Mobile Satellite Telephony Markets: 1999–2007*, the Asia-Pacific region will develop the largest base of subscribers with more than 30 percent of the global total. Most of them will be mobile users, although the fixed wireless market will grow dramatically with Latin America having 35 percent of the rural fixed wireless user base.

GEOSYNCHRONOUS SATELLITES

Clarke's concept of a stationary platform in space forms the basis for today's geostationary or geosynchronous satellites. Ringing the equator like a string of pearls, these birds provide a variety of services including 64-Kbps voice, broadcast television, video-on-demand services, broadcast and interactive data, and point-of-sale applications, to name a few. Although satellites are viewed as technological marvels, the real magic lies more in what it takes to harden them for the environment in which they must operate and what it takes to get them there than it does their actual operational responsibilities. Satellites are, in effect, nothing more than a sophisticated collection of assignable, on-demand repeaters—in a sense, the world's longest local loop.

From a broadcast perspective, satellite technology has a number of advantages. First, its one-to-many capabilities are unequaled. Information from a central point can be transmitted to a satellite in a geostationary orbit. The satellite can then rebroadcast the signal back to earth, covering an enormous service footprint. Consider, for example, TCI's *Headend in the Sky* (HITS) model. From the company's National Digital Television Center in Littleton, Colorado, TCI can generate a remarkable variety of programming content that is digitized, compressed, and uplinked to a constellation of satellites. The satellites then rebroadcast the content to every headend in the system, making it possible for the tiniest cable system in the company to offer the same channel lineup as the largest. Similarly, direct broadcast satellite companies like PrimeStar and DirecTV can offer equally robust channel lineups to their far-flung customers.

Because the satellites appear to be stationary, the earth stations actually *can* be. One of the most common implementations of geosynchronous technology is seen in the *Very Small Aperture Terminal* (VSAT) dishes that have sprung up like mushrooms on a summer lawn. These dishes provide both broadcast and interactive applications. The small DBS dishes that receive TV signals are examples of broadcast applications, while the dishes seen on the roofs of large retail establishments, automobile dealerships, and convenience stores are typically (although not always) used for interactive applications such as credit card verifications, inventory queries, e-mail, and other corporate communications. Some of these applications use a satellite downlink but rely on a telco return for the upstream traffic.

One disadvantage of geosynchronous satellites has to do with their orbital altitude. On the one hand, because they are so high, their service footprint is extremely large. On the other hand, because of the distance from the earth to the bird, the typical transit time for the signal to propagate from the ground to the satellite (or back) is about half a second, which is a significant propagation delay for many services. Should an error occur in the transmission stream during transmission, the need to detect the error, ask for a retransmission, and wait for the second copy to arrive could be catastrophic for delay-sensitive services like voice and video. Consequently, many of these systems rely on forward error correction transmission techniques that enable the receiver to not only detect the error, but correct it as well. The biggest supplier in the country for VSAT and other geosynchronous services is Hughes, although others exist, including GTE, SpaceNet, Inmarsat, Intelsat, and RCA.

Low/Medium Earth Orbit Satellites (LEO/MEO)

In addition to the geosynchronous satellite arrays, a variety of lower orbit constellations have been deployed, known as *Low and Medium Earth Orbit* (LEO/MEO) satellites. Unlike the geosynchronous satellites, these orbit at lower altitudes of 400 to 600 miles, far lower than the 23,000-mile altitude of the typical GEO bird. As a result of their lower altitude, the transit delay between an Earth station and a LEO satellite is virtually nonexistent. However, another problem exists with LEO technology. Because the satellites orbit pole to pole, they do not appear to be stationary, which means that if they are to provide uninterrupted service they must be able to hand off a transmission from one satellite to another before the first bird disappears below the horizon. This has resulted in the development of sophisticated satellite-based technology that emulates the functionality of a cellular telephone network. The difference is that in this case, the user does not appear to move; the cell does!

IRIDIUM

Perhaps the best known example of LEO technology is Motorola's ill-fated Iridium deployment. Comprising 66 satellites[18] in a polar array, Iridium was designed to provide voice service to any user anywhere on the face of the Earth. The satellites would provide global coverage and would hand off calls from one to another as the need arose. Unfortunately, Iridium's marketing strategy was flawed; their prices were high, their phones large and cumbersome (one newspaper article referred to them as "manly phones"), and their market significantly overestimated. Additionally, their system was only capable of supporting 64-Kbps voice services, a puny bandwidth allocation in these days of customers with broadband desires.

Iridium is not alone. ICO Global Communications, another satellite-based global communications company, filed for bankruptcy in August 1999. In November of that same year, Craig McCaw, Teledesic, and Eagle River offered salvation for the company with an investment package valued at $1.2 billion.

GLOBALSTAR

Others have been successful, however. Globalstar's 48-satellite array offers voice, short messaging, roaming, global positioning,

---

[18]The system was named Iridium because in the original design the system was to require 77 satellites, and 77 is the atomic number of that element. Shortly after naming it Iridium, however, the technologists in the company determined that they would only need 66 birds, but they did not appropriately rename the system Dysprosium.

fax, and data transport up to 9,600 bps. Although the data rates are minuscule by comparison to other services, the converged collection of services they provide is attractive to customers who want to reduce the number of devices they must carry with them in order to stay connected.

ORBCOMM

ORBCOMM is a partnership jointly owned by Orbital Sciences Corporation and Teleglobe Canada. Their satellites are nothing more than extraterrestrial routers that interconnect vehicles and earth stations to facilitate the deployment of such packet-based applications as two-way short messaging, e-mail, and vehicle tracking. Their constellation has a total of 35 satellites.

TELEDESIC

After Iridium, the best-known LEO satellite services company is Teledesic. Started in 1990 by Craig McCaw and Bill Gates, Teledesic is comprised of an array of 288 satellites capable of providing up to 64 Mbps downstream and 2 Mbps upstream for voice, videoconferencing, and data, with plans in place to offer a symmetric 64-Mbps service in the future. In 1998, Motorola joined the Teledesic team, and in 1999, the company signed a launch agreement with Lockheed Martin. They plan to be fully operational by 2004.

IN SUMMARY

As a service-provisioning technology, satellites may seem so far out (no pun intended) that they may not appear to pose a threat to more traditional telecommunications solutions. At one time, that is, before the advent of LEO technology, this was largely true. GEO satellites were extremely expensive, offered low bit rates, and suffered from serious latencies that were unacceptable for many applications.

Today this is no longer true. Some GEO satellites offer high-quality, two-way transmissions for certain applications. LEO technology has advanced to the point that it now offers low-latency, two-way communications at broadband speeds, is relatively inexpensive, and, as a consequence, poses a clear threat to terrestrial services. On the other hand, the best way to eliminate an enemy is to make the enemy a friend. Many traditional service providers have entered into alliances with satellite providers. Consider the agreements that exist between DirecTV, a high-quality, wireless alternative to cable, and Bell Atlantic, GTE, Cincinnati Bell, and the SBC corporations. Again, convergence is at work here. Customers have indicated repeatedly that they like the idea of being able to go to a single source for all of their communications needs. By joining forces with satellite providers, the ILECs create something of a market block that will help them stave off the incursion of cable. Between the minimal infrastructure required to receive satellite signals and the soon-to-be ubiquitous deployment of DSL over twisted pair, incumbent local telephone companies and their alliance partners are in a reasonably good position to counter the efforts of cable providers wanting to enter the local services marketplace.

THE SPECIAL CASE OF PREMISES ACCESS

It would be improper to discuss access technologies without also discussing the technologies used to *access* the access technologies—LANs and variations on the LAN theme. LANs have traditionally fallen into two primary categories characterized by the manner in which they access the shared transmission medium (shared among all the devices on the LAN). The first, and most common, is called *contention,* and the second group is called *distributed polling.*

CONTENTION-BASED LANS

In contention-based LANs, devices attached to the network vie for access using the technological equivalent of gladiatorial combat. "If it feels good, do it" is a good way to describe the manner in which they share access. If a station wants to transmit, it simply does so, knowing that there exists the possibility that the transmitted signal may collide with that generated by another station that transmits at the same time. In the event that a collision occurs, both stations back off, wait a random amount of time, and try again. Ultimately, each station *will* get its turn, although how long they have to wait is based upon how busy the LAN is. These systems are characterized by what is known as *unbounded delay*, because there is no upward limit on how much delay a station can incur as it waits to use the shared medium.

The protocol that these LANs usually employ is called *Carrier Sense, Multiple Access with Collision Detection* (CSMA/CD). In CSMA/CD, stations observe the following guidelines when attempting to use the shared network. First, they listen to the shared medium to determine whether it is in use or not; that's the Carrier Sense part. If the LAN is available, they begin to transmit but continue to listen while they are transmitting, knowing that another station could also choose to transmit at the same time; that's the Multiple Access part. In the event that a collision is detected, usually indicated by a dramatic increase in the signal power measured on the shared LAN, both stations back off and try again; that's the Collision Detection part.

Ethernet is the most common example of a CSMA/CD LAN. Originally released as a 10-Mbps product based on IEEE standard 802.3, Ethernet rapidly became the most widely deployed LAN technology in the world. As bandwidth-hungry applications such as e-commerce, *Enterprise Resource Planning* (ERP), and Web access evolved, transport technologies advanced, and bandwidth availability (and capability) grew, 10-Mbps Ethernet began to show its age.

The other aspect of the LAN environment that began to show weaknesses was the overall topology of the network itself. LANs are broadcast environments, which means that when a station transmits, every station on the LAN segment hears the message. Although this is a simple implementation scheme, it is also wasteful of bandwidth, since stations hear broadcasts that they have no reason to hear.

In response to this, a technological evolution occurred. It was obvious to LAN implementers that the traffic on most LANs was somewhat domain-oriented. That is, it tended to cluster into communities of interest based upon the work groups using the LAN. For example, if employees in Sales shared a LAN with Shipping and Order Processing, three discernible traffic groupings emerged according to what network architects call the *80:20 Rule* (see Figure 1-15). The 80:20 Rule simply states that 80 percent of the traffic that originates in a particular work group tends to stay in that work group, an observation that makes network design distinctly simple. If the traffic naturally tends to segregate itself into groupings, then the topology of the network could change to reflect those groupings. Thus was born the bridge.

Bridges are devices with two responsibilities: they filter traffic that does not have to propagate in the forward direction and forward traffic that does. For example, if the network described earlier were to have a bridge inserted in it, three work groups would each have their own LAN segment, and each segment would be attached to a port on the bridge (see Figure 1-16). When an employee in Sales transmits a message to another employee in Sales, the bridge is intelligent enough to know that the traffic does not have to be forwarded to the other ports. Similarly, if the Sales employee sends a message to someone in Shipping, the bridge recognizes that the sender and receiver are on different segments and thus forwards the message to the

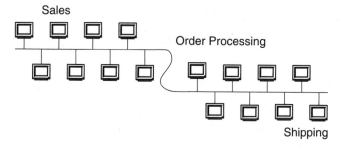

FIGURE 1-15   Three departments, a single LAN

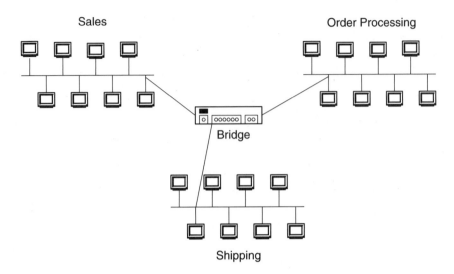

FIGURE 1-16   Three departments, a single LAN segmented with a bridge

appropriate port, using address information in a table that it maintains (the filter/forward database).

This technique is not unique, nor is it the only solution possible. Network designers have long relied on hubs to create a star-wired scheme in which all "LAN loops" (often unshielded twisted pair, or UTP) radiate outward from a central device. This provides the network administrator with a single location from which the network can be administered and managed.

Following close on the heels of the hub concept is a relatively new technique called *LAN switching*. LAN switching qualifies as "bridging with an attitude." In LAN switching, the filter/forward database is distributed; that is, a copy of it exists at each port, which implies that simultaneous traffic-handling decisions can be made by different ports. This also enables the LAN switch to implement full-duplex transmission, reduce overall throughput delay, and in some cases implement per-port rate adjustments. The first 10-Mbps Ethernet LAN switches emerged in 1993, followed closely by Fast Ethernet (100-Mbps) versions in 1995 and Gigabit Ethernet (1,000-Mbps) switches in 1997. Fast Ethernet immediately stepped up to the marketplace bandwidth challenge and was quickly accepted as the next generation of Ethernet, with per-port prices dropping 47 percent between 1997 and 1998 in concert with a four-fold increase in port sales during the same timeframe.

Gigabit Ethernet is still in a somewhat nascent stage, but most believe that it will experience a similar uptake rate. Dataquest predicts that Gigabit Ethernet sales will grow to $2.5 billion by 2002. This is a reasonable number, considering that two million ports were sold in 1999 with expectations of hitting a total installed base of 18 million by 2002. Emerging applications certainly make the case for Gigabit Ethernet's bandwidth capability. LAN telephony, server interconnections, and video to the desktop all demand low-latency solutions, and Gigabit Ethernet may be positioned to provide it. A number of vendors have entered the marketplace including Alcatel, Lucent Technologies, Nortel Networks, Fore Systems, Cisco Systems, and Cabletron.

DISTRIBUTED POLLING LANs

In addition to the gladiatorial combat approach to sharing access to a transmission facility, a more civilized technique can be utilized, known as *distributed polling*, or as it is more commonly known, *token passing*. IBM's Token-Passing Ring is perhaps the best known of these products, followed closely by *Fiber Distributed Data Interface* (FDDI), a 100-Mbps version often seen in campus and *metropolitan area networks* (MANs).

In token-passing LANs, stations take turns with the shared medium, passing the right to use it from station to station by handing off a "token" that gives the bearer the one-time right to transmit while all other stations remain quiescent. This is a much more fair way of sharing access to the transmission medium than CSMA/CD, because, although every station has to wait its turn, it is absolutely guaranteed that it will get that turn. These systems are therefore characterized by bounded delay, because there is a maximum amount of time that any station will ever have to wait for the token.

Traditional token ring LANs operate at two speeds: 4 and 16 Mbps. Like Ethernet, these speeds were fine for the limited requirements of text-based LAN traffic that was characteristic of early LAN deployments. However, as demand for bandwidth climbed, the need to eliminate the bottleneck in the token ring domain emerged and fast token ring was born. In 1998, the IEEE 802.5 committee (the oversight committee for token ring technology) announced draft standards for 100-Mbps, high-speed token ring (HSTR 802.5t). A number of vendors stepped up to the challenge and began to produce high-speed token ring equipment including Madge Networks, IBM, and Olicom (which has since sold its token ring business to Madge).

Gigabit token ring is on the horizon as draft standard 802.5v, and although it has not yet emerged as a widespread access technology, it undoubtedly will as demand for bandwidth grows. However, token ring tends to be more expensive than Ethernet, and given the widespread deployment and over-whelming global success of fast and gigabit Ethernet, high-speed token ring may have a battle before it.

THE HOME PHONELINE NETWORKING ALLIANCE (HOMEPNA)

The *Home Phoneline Networking Alliance* (HomePNA) is a not-for-profit association founded in 1998 by 11 companies to bring about the creation of uniform standards for the deployment of interoperable in-home networking solutions. The companies, 3Com, AMD, AT&T Wireless, Compaq, Conexant, Epigram, Hewlett-Packard, IBM, Intel, Lucent Technologies, and Tut Systems, are committed to the design of in-home communications systems that will enable the transport of LAN protocols across standard telephone "inside wire."

Currently, HomePNA operates at one Mbps and enables the user to employ every RJ-11 jack in the house as both telephone and data ports for the interconnection of PCs, peripherals, and telephones. The standard enables interoperability with existing access technologies such as xDSL and ISDN and therefore provides a good migration strategy for *Small Office/Home Office* (SOHO) and telecommuter applications. The initial deployment utilizes Tut Systems' HomeRun product, which enables telephones and computers to share the same wiring and to simultaneously transmit. Extensive tests have been performed by the HomePNA, and although some minor noise problems caused by AC power noise and telephony-coupled impedance have resulted, they have been able to eliminate the effect through the use of low-pass filters between the devices causing the noise and the jack into which they are plugged.

ACCESS TECHNOLOGIES SUMMARY

This is a book about convergence and as such we must consider how the convergence phenomenon affects the access technologies discussed in this section. We have examined a variety of options for connecting the customer to the wide area transport network including ISDN, T-1, cable modems, 56K modems, DSL, wireless solutions such as LMDS, MMDS, and satellites, and premises schemes such as LANs and HomePNA technology. In all of these, we have seen one recurring theme: the evolution of technology in each case to satisfy the demand for network bandwidth that will meet the needs of ever-evolving applications required by the customer to achieve a competitive advantage in their marketplace.

Remember the convergence mantra: Discrete knowledge about the customer's business goals and activities translates into a choice of technologies. The customer will ask for the best access solution possible; the access provider must respond with the appropriate collection of technologies and must not burden the customer with how those underlying technologies work. Ideally, what the customer sees is not the bits and bytes of technology, but rather a solution to a business problem.

Each of the access technologies described fits a particular niche. The key to success lies in correctly matching a unique technology to a specific solution. ISDN, for example, is the target of criticism and even ridicule in some circles, but it *does* offer 128-Kbps service *right now* in the areas where it is available, something 56-Kbps modems can't begin to achieve. Although DSL seems like the ideal technology for most problems, it is far from universally available and still suffers from vexing problems that stymie its widespread deployment in the near-term. DSL will certainly overcome its problems, but in the meantime cable modems stand ready to fill the broadband access void with their own flavor of high-speed access as well as content. Unfortunately, cable modems face the same spotty deployment challenges that plague the DSL providers, not to mention that cable is nowhere near as widely deployed as twisted pair. There is no question that the ILECs have a significant advantage with their switched local loop plant, but companies like AT&T are closing the gap by forming alliances with cable providers like TCI. They understand that the magic formula in the game has more than one ingredient. Remember the words of Claudio Monasterio of Lucent Technologies: "The technology is nothing more than a detail."

Service providers must take heed of their title and role in the converging marketplace. They are and are expected to be *service providers*. The difficult part of that expectation is the required evolution from being an access and transport provider to being a provider of customer solutions based on the technologies they understand, manage, and sell. As we will see in a later section, service providers must learn to focus not exclusively on their immediate customer, but on the needs of their *customer's customer*. After all, the immediate customer will use whatever solution is provided to them to create solutions to their own customers' business challenges. If service providers take the initiative to look beyond their immediate market to that third tier, make an effort to understand the issues faced there, and craft solutions that will satisfy both the second- and third-tier business concerns, they win the game. This is not a game of who has the best technology; it is a game of who provides the best service. If technology happens to be the basis for the service, so be it. The customer, however, shouldn't have to care.

TRANSPORT TECHNOLOGIES

Transport technologies, usually represented as the famous cloud seen more or less ubiquitously in telecommunications product and service literature, comprise both dedicated private-line facilities and switched technologies, all designed to move traffic from one access point to another at the highest speed possible.

Traditionally, private-line facilities were the most common technique, with bandwidth ranging from as little as 300 bits per second in the early days of data transmission to the almost unimaginable gigabit speeds of the North American *Synchronous Optical Network* (SONET) and its global counterpart, the *Synchronous Digital Hierarchy* (SDH). Add to those the remarkable bandwidth multiplication capabilities provided by *Wave Division Multiplexing* (WDM), and we soon find that bandwidth is no longer the bottleneck that it once was. In fact, there is so much bandwidth available today that for the first time, commodity markets are appearing where it can be purchased like any other commodity—soy beans, pork bellies, Louisiana sweet crude, and bandwidth.

This poses a threat to those companies that have traditionally made their profits through the sale of bandwidth as a limited resource. As companies like Qwest and Level 3 build out their optical networks and bill themselves as the "carriers' carriers," a once-scarce resource now becomes so widely available that the price for it plummets to near zero. When a product becomes so widely available from multiple suppliers that the only way to differentiate between them is on price alone, the product officially becomes a commodity. Obviously, service providers do not want to play in the commodity game, but if they view themselves as nothing more than access and transport providers, that's what they will become. Keep in mind, however, that the *good* news about being a commodity provider is that by definition everyone wants your product. To maintain a competitive advantage, service providers must heed the words of Harvard business professor Michael Porter, author of numerous books on competitive advantage and strategy: "A firm differentiates itself from its competitors when it provides something unique that is valuable to buyers beyond simply offering a low price."[19] In the access and transport game, everyone has a network, and by and large, they are all equally capable. Therefore, having the best network or the most expensive technology does

---

[19]Porter, Michael. *Competitive Advantage: Creating and Sustaining Superior Performance.* New York; The Free Press, 1985, page 120.

not give service providers an advantage, because the network itself is a commodity. Traditional access and transport companies must therefore become more than merely a data transport company. They must overlay a set of services on their physical networks that change them from commodity providers into *service* providers. At that moment, the underlying technology becomes a valuable piece of the overall package.

During the early 1980s, when voice was the predominant traffic component on service provider networks, it was occasionally stated that "if we build a voice network, the data can ride for free." This observation made perfect sense. Data transmission at that time represented a tiny component of the overall traffic volume. Not only that, it soon became a fundamental component of the competitive voice services market. With the implementation of the SS7 data network and its dedicated databases, it soon became possible to offer so-called intelligent network applications, which became enormous moneymakers for service providers.

Consider the following example. A customer with a flat rate monthly service receives a telephone call but doesn't get to the phone before the caller hangs up. The customer examines their Caller ID display (for which he pays an additional monthly fee) and decides he wants to speak with the caller. They dial *69, which invokes an SS7 service that automatically calls the caller back for an additional fee. While they are on the phone, the customer hears a tone in his or her ear, indicating another incoming call that they then answer while putting the other person on hold, all for a fee. At some point during the call, they add on a third party using three-way calling, again for a fee.

Please note that in this scenario, no charge is associated with the duration of the call itself other than the low, flat rate monthly fee. However, this simple phone call, because of the added data services that SS7 provides, generates approximately $6 in added revenue. By adding data services on top of the voice network's simple transport capabilities, *data becomes the killer application for voice.*

Enter the 1990s. Voice is no longer the predominant traffic component on modern networks, yet it still accounts for the majority of the revenue. In August 1998, SBC Executive Vice President for Corporate Planning Michael Turner observed that

98 percent of the company's revenue derived from circuit-switched voice communications and went on to observe that even when data becomes more than half of the traffic SBC transports across their network, it will still account for less than half of their revenue. Even today, voice is the major source of revenue, as expressed in this quote from Jim Crowe, founder and CEO of Level 3 Communications:[20]

> "Data is, by bit count, some 50 percent of the traffic flow through networks. Yet voice still represents 91 cents of each dollar billed to customers because telecommunications companies charge 15 times more for data than they do for voice. So there's a huge opportunity to profit based on the telecom market's shift to carrying data. But voice's revenue-generating advantage in billing will not disappear overnight. So beware of providers who concentrate on the multimedia data market and ignore the voice market. They could go broke if the multimedia market fails to grow quickly enough."

So although data turned out to be the killer application for voice networks, service providers are now faced with the challenge of determining what the killer application for data networks will be as the world migrates to a converged access and transport infrastructure. Many believe, ironically, that the killer application for data networks will be voice.

CLOUDS

So why do we draw clouds to represent the transport network? The typical answers are

· We don't know what's in there.
· We don't *want* to know what's in there.
· There's no *need* to know what's in there.

---

[20]*Telecom Shakeup Offers IT Opportunities.* IEEE ITPro, May/June 1999.

- It doesn't *matter* what's in there.
- Clouds are easier to draw than hairballs.

To a large extent, all of the aforementioned reasons are true. In many cases, we *don't* know what's in the cloud because we've never had to know. In the days prior to divestiture, when there was only one game in town from which to choose (called AT&T), the service selection scenario might have gone something like this:

Customer: "Look, I've got a big problem. I'm a large bank with branches all over town. I have a mainframe computer in the data center downtown, I've got tellers in all the branches with dumb terminals that have to talk with it, I've got a few ATM machine scattered here and there, and—" (interrupted)

AT&T Salesperson: "Shhhh . . . look, don't worry about the stuff in the middle. You just tell us where each location is and what they do there, and we'll take care of the stuff in the middle. *You take care of your banking operations and let us take care of the network—that's our job.*"

In this scenario, the network sales person did not really know or *want* to know what was in the cloud, nor did the customer. The important thing is that there was no *need* for either one to know because there were plenty of highly technical people in the phone company's ranks who would see to it that the bank got what they needed for continuing business operations. The AT&T person was free to concentrate on the customer's business issues.

That takes care of the first three bullets in the list. The fourth, *It doesn't matter what's in the cloud,* is also true. In the ideal world of service provisioning, the customer should have no reason to care what's in there. All they should have to care about is that they are given a "gozinta" and a "gozouta" that disappear into the fluffy opaqueness of the cloud, and as long as they connect to them correctly, their voice and data needs will be met. End of story.

This scenario represents the ideal world of service provisioning. If providers build their networks with the customers' requirements in mind (knowledge) and therefore choose the right technologies, they can sell access to a multipurpose cloud that satisfies *all* of the customer's network requirements. They become the provider of choice because there is no reason for the customer to go anywhere else. The service provider delivers exactly what the customer requires as well as where and when they need it. Furthermore, they may form strategic alliances with providers of other desirable services such as content, as demonstrated by the relationships that exist between AOL, Prodigy, and the ILECs. They are no longer merely access and transport companies; they are true *service* providers.

THE EVOLUTION OF TRANSPORT

In this section, we will examine the principal transport technologies, beginning with dedicated facilities such as private line, SONET/SDH, and microwave. Then we will continue with the evolution of switched services from traditional circuit and packet switching to modern fast packet architectures. The section will conclude with routing and the special case of the IP, the Rosetta Stone of communications technologies.

POINT-TO-POINT ARCHITECTURES

Point-to-point technologies do exactly what their name implies. They connect one point directly with another. For example, it is common for two buildings in a downtown area to be connected by a point-to-point microwave or infrared circuit, because the cost of establishing it is far lower than the cost to put in physical facilities in a crowded city. Many businesses rely on dedicated, point-to-point optical facilities to interconnect locations, especially businesses that require dedicated bandwidth for high-speed applications. Of course, point-to-point does not necessarily imply high-bandwidth. Many locations use 1.544-Mbps T-1 facilities for interconnection, and some rely on lower-speed circuits where higher bandwidth is not required.

Dedicated facilities provide bandwidth from as low as 2,400 bits per second to as high as multiple gigabits per second. 2,400-bps analog facilities are not commonly seen but are often used for alarm circuits and telemetry, while circuits operating at 4,800 and 9,600 bps and used to access interactive, host-based data applications.

Higher-speed facilities are usually digital and are often channelized by dedicated multiplexers and shared among a collection of users or by a variety of applications. For example, a high-bandwidth facility that interconnects two corporate locations might be dynamically subdivided into various-sized channels for use by a PBX for voice, a videoconferencing system, and data traffic.

Needless to say, satellite service, LMDS, and MMDS are also potential candidates for transport. They have already been discussed in detail in the prior section but should be considered if the opportunity to use them arises.

HIGH-BANDWIDTH POINT-TO-POINT SERVICES

HIGH-BANDWIDTH WIRELESS SERVICES. Wireless services include microwave, which can provide connectivity at speeds in the gigabit-per-second range, and infrared, which typically operates in the tens of megabits per second range. The advantages of both include a high transmission speed and relatively low deployment costs. They also include disadvantages. Both require a line-of-sight transmission, which means that each end of the circuit must have an unimpeded view of the other end. In a microwave transmission, the signal can be repeated (a typical microwave hop is about 30 miles), but this adds cost and complexity to the deployed circuit. Furthermore, microwave is susceptible to atmospheric interference from fog and rain (often called *rain fade*), which lowers its effectiveness in certain geographical areas. The technology also requires an operational license because of the frequency in which it operates and must be monitored carefully to avoid interference with other radio-based services.

Infrared is used for shorter hops because it is extremely secure, operates at a high transmission speed, and requires no operating license. It can be used both inside and outside of a building. It is often deployed in secure facilities, because it will not propagate outside of the building. Some radio-based systems emit energy beyond the confines of the building in which they are deployed, making their use less than desirable when sensitive information is being transmitted.

Infrared generally does not offer the distance nor the bandwidth that microwave can provide, but it is less expensive. In the local area environment, infrared can easily operate at 100 Mbps with extremely low error rates. In the wide area, it can operate up to 34 Mbps but often provides lower throughout because of atmospheric interference.

HIGH-BANDWIDTH WIRED SERVICES. In the copper realm, DS-1 (usually called T-1) and DS-3 are the most commonly deployed services for high-bandwidth applications. T-1 operates at 1.544 Mbps, while DS-3 is the equivalent of 28 DS-1s, offering a respectable 44.736 Mbps. Both can be sold as either a channelized or unchannelized service, as can their international counterparts E-1 (2.048 Mbps) and E-3 (usually 34 Mbps).

In North America, the standard digital multiplexing hierarchy that begins with DS-0 effectively ends at DS-3. There have always been multiplexers available operating at speeds in excess of DS-3's 45 Mbps, but until the advent of the SONET, those devices used proprietary framing schemes that limited their usefulness due to a lack of interoperability with different vendors' devices.

OPTICAL SYSTEMS

Optical transmission has a number of significant advantages over copper-based media. First, it is immune to the electromagnetic interference that plagues metallic media and can therefore be installed with less concern for its proximity to noise-generating devices such as electric motors, fluorescent lights, and so on. It is also difficult, although not impossible, to tap and is therefore more secure than copper facilities. It does not require regeneration equipment placed as frequently in the circuit as metallic circuits do, because optical signals do not weaken as quickly as electrical signals. Finally, optical technology provides unprecedented bandwidth, terabits per second in some cases.

SYNCHRONOUS OPTICAL NETWORK (SONET). SONET is an internationally recognized multiplexing standard that begins where DS-3 leaves off. Proposed initially in 1984, SONET was proposed as a solution to the problem of vendor interoperability following the divestiture of AT&T. Integral to the breakup was the concept of equal access, designed to ensure that customers had the right to choose their long-distance carrier from among (at the time) the big three: AT&T, MCI, and Sprint. The problem was that most central offices in the country were replete with Western Electric (AT&T) equipment, which meant that if MCI or Sprint wanted to interconnect with the customers (and therefore equipment) served out of a former AT&T central office, they were obligated to purchase Western Electric hardware to ensure interoperability, since there were no optical interconnect standards at the time. Clearly this did not sit well with MCI and Sprint, who did not want to be obligated to one vendor or another, hence the arrival and rapid success of the SONET standard.

SONET's lowest transport speed, called *Optical Carrier Level One* (OC-1) is 51.84 Mbps. Higher speeds can be accommodated by allocating even multiples of OC-1 in a series of recognized combinations, as shown in Table 1-3.

In the same way that T-Carrier systems define both a framing standard (T-1) and a multiplexing hierarchy (DS-1), SONET defines both an *Optical Carrier level* (OC) and its electrical equivalent, known as the *Synchronous Transport Signal* (STS). Unlike T-Carrier, the OC/STS levels are exact multiples of the base rate. For example, the bandwidth of a T-1 (1.544 Mbps) is not equal to 24 64-Kbps channels because of the added overhead for framing. Similarly, a DS-3's bandwidth is significantly higher than that of 28 DS-1s for the same reasons. In SONET, however, an OC-12 is *exactly* 12 times the bandwidth of an OC-1. This enables SONET (or SDH) to easily and transparently carry payloads at any conceivable speed. A DS-3 fits rather nicely in an OC-1 with a little room left over. A 100-Mbps Fast Ethernet stream can easily be accommodated within a SONET OC-3, with a little left over. SONET operates under the belief that bandwidth is cheap, and it's a good thing because it wastes an awful lot of it in unused bytes.

Like T-Carrier, SONET can be configured as either a channelized or unchannelized service. If it is channelized, multiple OC-1s are multiplexed together on a common fiber. For example, an OC-12 provides 622.08 Mbps of bandwidth, but the fastest speed attainable from the service is OC-1. The user could derive 12 of them, but they would all operate at the fundamental OC-1 rate. T-Carrier works in the same way. The user can get 24 64-Kbps channels or a single "pipe" operating at 1.536 Mbps.

TABLE 1-3 Acclerating OC-1

OPTICAL CARRIER LEVEL	ELECTRICAL LEVEL	BANDWIDTH	SDH STM EQUIVALENT
OC-1	STS-1	51.84 Mbps	—
OC-3	STS-3	155.52 Mbps	STM-1
OC-9	STS-9	466.560 Mbps	STM-3
OC-12	STS-12	622.08 Mbps	STM-4
OC-18	STS-18	933.120 Mbps	STM-6
OC-24	STS-24	1244.16 Mbps	STM-8
OC-36	STS-36	1866.24 Mbps	STM-13
OC-48	STS-48	2488.32 Mbps	STM-16
OC-96	STS-96	4976.64 Mbps	STM-32
OC-192	STS-192	9953.28 Mbps	STM-64

If more bandwidth is required for a higher speed application, the channels can be combined. In this case, our OC-12 becomes an OC-12c, where the c stands for *concatenated*, which means "chained together." Here we create a service that is analogous to an unchannelized T-Carrier, where the user is provided with a single circuit that operates at the aggregate rate of the entire facility.

SONET is also fully capable of transporting smaller payloads (ones that require less bandwidth than an OC-1 provides, such as T-1). In this case, the OC-1 is broken into smaller payload components called *virtual tributaries*, each of which is capable of transporting subrate services. Table 1-4 displays the four identified virtual tributary types with the payload they are each capable of conveniently transporting.

OTHER SONET ADVANTAGES. Besides providing interoperability (also known as "midspan meet," referring to the capability of two multiplexers from different manufacturers to "meet" in the middle of an optical span and transparently, no pun intended, swap data), SONET offers a number of other advantages as well. First, it provides a standard multiplexing scheme for services that require bandwidth in excess of DS-3. Second, it provides embedded, well-designed network management capabilities. Third, it dramatically simplifies the process of adding and dropping ("groom and fill") payload components along the path. Fourth, it facilitates multipoint configurations, enabling payload to be added and dropped at any point in the network with a minimal amount of additional complexity. And finally, SONET provides for enormous bandwidth, making it possible to transport essentially any payload presented to it.

TABLE 1-4   Virtual Tributaries

VIRTUAL TRIBUTARY TYPE	PAYLOAD TYPE	BANDWIDTH
VT1.5	DS-1	1.728 Mbps
VT2	E-1	2.304 Mbps
VT3	DS-1C	3.456 Mbps
VT6	DS-2	6.912 Mbps

SONET ARCHITECTURES. **SONET** networks are usually deployed in dual-ring configurations, as shown in Figure 1-17. In other words, all elements along the span are connected to dual counter-rotating optical rings in which one serves as the primary path, the other as a backup. In this case, the elements on the ring are add-drop multiplexers that have the capability to add and drop payload easily. They are simultaneously connected to both the primary and secondary rings. In the event that the primary ring fails, the devices along the failed path will detect the failure and wrap, as shown in Figure 1-18, sealing the ring within 50 ms and preventing a catastrophic failure of the ring. This capability, known as *Automatic Protection Switching* (APS), is one of the key advantages of the SONET ring architecture. It is routinely deployed in major networks to eliminate the possibility of a total failure of the network, something that traditional private lines cannot accomplish.

Wavelength Division Multiplexing (WDM). The **OC-192** (9.9 Gbps) transmission rate that SONET delivers is staggering, yet in many cases is not enough for today's graphic and video-intensive applications. Before the arrival of the Internet and its World Wide Web application, the idea that we would ever need such enormous amounts of transport capacity was ludicrous. However, Parkinson's Law is proving to be true as much in telecommunications as it is in our personal lives. The original iteration of Parkinson's Law stated that "Work expands to fill the time allotted to it." A second variation observed that "Expenditures will always rise to exceed income." Today Parkinson's Law tells us that there is no such thing as a fast enough processor, enough installed memory, a big enough hard drive, or too much bandwidth. Traditional optical fiber is proving to be inadequate for current bandwidth demand.

In response, two solutions have emerged. The first is to simply deploy more fiber, which solves the problem, but at enormous expense, as much as $70,000 per mile in some cases. Alternatively, a relatively new technology has emerged called *Wavelength Division Multiplexing* (WDM), discussed earlier. Capable of multiplying the overall capacity of a single fiber several times over, WDM lowers the cost of deployed bandwidth from $70,000 per mile to as little as $20,000, because the change

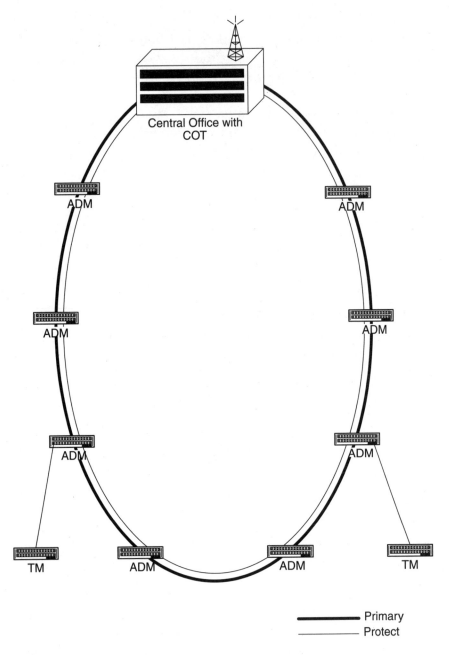

FIGURE 1-17   SONET dual-ring configuration

requires no backhoe. A simple replacement of endpoint electronics takes care of the conversion.

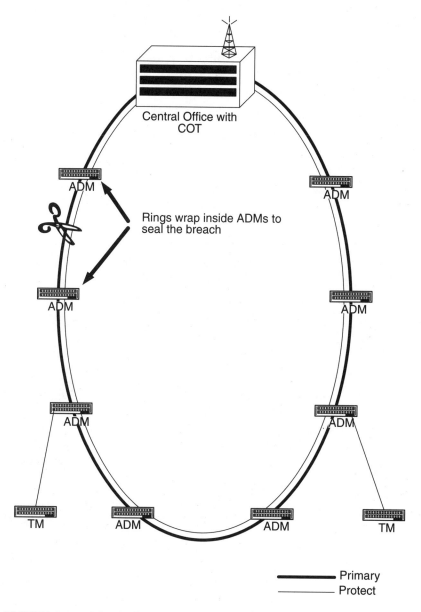

Central Office with
COT

Rings wrap inside ADMs to
seal the breach

ADM

ADM

ADM

ADM

ADM

ADM

TM

ADM

ADM

TM

———— Primary
———— Protect

FIGURE 1-18   SONET dual-ring configuration when the primary ring is breached

HOW IT WORKS.   In optical systems, light from a high-quality laser (sometimes a *light emitting diode* [LED]) is transmitted down the fiber to a receiver, which converts the optical energy to an electrical signal so that it can be processed. The light transmitted is in the infrared range of the electromagnetic spectrum and is therefore not visible. A typical fiber-based system without WDM transmits a single information stream at high speed, typically as high as 2.4 Gbps (OC-48). With WDM technology, that same fiber can carry multiple streams of information at high speeds, thus multiplying the actual capacity of each fiber.

The original WDM systems carried four individual channels, while modern systems can carry as many as 16 or more. Experimentation has shown that it is possible to build systems capable of transporting as many as 1,000 distinct channels, although it will be some time before they become commercially feasible. A number of vendors now sell multi-channel systems. Lucent, Cambrian, Ciena, Corvis, and Fujitsu, to name a few, have demonstrated high-capacity systems with as many as 160 supported channels. In November 1999, Lucent announced its WaveStar™ LambdaRouter, which supports 256 channels, each operating at 40 Gbps. Others will certainly follow with even more densely packed channels.

Network implementers themselves have stepped up to support *Dense Wavelength Division Multiplexing* (DWDM) technology as well, also discussed earlier. MCIWorldcom/Sprint and AT&T have announced not only domestic augmentations to their fiber networks, but international alliances as well. AT&T has entered into agreements with British Telecom (at the expense of its relationship with Unisource) and a number of other providers in Europe and Latin America, while MCIWorldcom has alliances with Telefónica of Spain and Embratel of Brazil.

Among the ranks of the so-called "bandwidth barons," Level 3 and Qwest have made enormous strides with the deployment of their own networks. Level 3, which started life as part of a construction company (the same company that gave birth to Metropolitan Fiber Systems), is now building a 15,000-mile fiber network over which it plans to run IP throughout North America. They acquired softswitch manufacturer Xcom, which integrates circuit and packet-based traffic for IP transport. Qwest, spun off from Southern Pacific Railroad in 1988, has built an 18,500-mile network and is now the fourth largest (counting Sprint and

MCIWorldcom as two companies) long-distance company in the U.S. The firm also has a significant presence in both Asia and Europe through a series of alliances and acquisitions.

The fiber manufacturers themselves have also done a great deal to increase the distance and carrying capacity that their products support. Corning's LEAF fiber, introduced in February 1998, does not need regenerators as often as other fibers and requires less power to create a transportable signal. Most fiber is optimized for regeneration every 6 miles or so. LEAF only requires regenerators every 30 miles.

FIBER TRANSMISSION ISSUES AND CONCERNS. Although optical fiber offers all the advantages cited earlier—immunity from noise and interference, high security, enormous bandwidth, distance—it does have some disadvantages as well. First, although it is not affected by electromagnetic interference, optical interference problems can occur within the fiber, limiting the overall bandwidth of the facility. These include physical limitations of the electronics, distance-dependent signal degradation, *Stimulated Raman Scattering* (SRS), *Stimulated Brouillian Scattering* (SBS), cross-wave modulation, and four wave mixing. We will discuss each of these in turn.

First is the limitations of the electronics. Simply stated, lasers take a certain amount of time to cycle between different states (the so-called "rise time"), and the degree to which that period of time can be minimized is finite. There is therefore a physical limitation based on solid state electronics as we know them today.

Another optical interference problem is *distance-dependent signal degradation.* Even optical fiber suffers from signal degradation due to a number of factors such as chromatic dispersion, pulse spreading over distance, and a variety of others that will be discussed shortly. It is therefore necessary to install regenerators along optical spans that receive the incoming optical signal, convert it to electrical, clean and regenerate the original signal, and convert it back to optical for retransmission. An innovative technology has emerged in recent years that enables an optical amplification of the signal (that is, the signal does not have to be converted back to electrical, amplified, and reconverted to optical again). It is known as the *Erbium Doped Fiber*

Amplifier (EDFA). In EDFA systems, a segment of fiber with erbium atoms mixed into its silica matrix (a process called *doping*) is inserted in the optical strand. The EDFA segment is connected to optical electronics that have the capability to pump energy into selected frequencies, specifically those frequencies used to transport the actual content (between 1,480 and 1,600 nm). When the signal arrives at the EDFA segment, the incoming light pulse excites the erbium atoms embedded in the EDFA. (Sorry, a little physics is necessary here to illustrate the magic of this technology.) When the erbium atoms get excited, they jump up to the next quantum energy level, and when they fall back down again, they give off a photon with an energy level that happens to fall within the 1,480 to 1,600 nm range used by optical systems.[21] The resulting photons pump energy into the passing signal, effectively amplifying it along the way.

Stimulated Raman Scattering (SRS) is a phenomenon that typically affects large systems and occurs when high-frequency channels in a DWDM system pump energy into low-frequency channels, resulting in high-frequency signal loss. Short-wavelength, high-energy channels are the most depleted by this phenomenon, and the only way to avoid it is to operate the system at low power levels. This problem is also a benefit. SRS is precisely the phenomenon used in Erbium-doped fiber amplifiers!

Stimulated Brouillian Scattering (SBS) affects smaller systems, and only occurs when adjacent channels operate in opposite directions. In SBS, optical energy is scattered back toward the transmitter, resulting in a loss of signal power. Again, keeping the system below a predefined power threshold is the only way to eliminate SBS.

Cross Phase Modulation is a form of crosstalk that is the result of the fact that the refractive index of fiber changes with the intensity of the signal being propagated. As the intensity of the signal changes, so too does the phase shift that naturally occurs during transmission. The phase shift results in crosstalk,

---

[21]This range was not chosen randomly. At 1,480 and 1,550 nm, optical signals are least affected by the various physical impairments that weaken or disperse them, such as hydroxyl (water) absorption.

which like so many other impairments can only be eliminated by reducing the overall power of the transmitted signal.

Four Wave Mixing occurs when the frequency spacing between channels is so small that the different waves phase-lock. This results in the creation of sideband channels that carry no information but tend to bleed power out of the primary signals. The solution is to use *non-zero dispersion shifted fiber* (NZDSF), which was designed to enable a small amount of signal dispersion without the zero dispersion point falling within the WDM passband. Examples of NZDSF include Lucent's Truewave and Corning's LS fiber and LEAF fiber.

A WORD ABOUT OPTICAL SWITCHING. It is now possible to build an all-optical switch, at least on a small scale. The most common channel separation technique is to use a device called a *Bragg Grating Filter*. The Bragg Grating is nothing more than a segment of fiber that has been treated in such a way that it becomes a highly specific diffraction grating that reflects a single wavelength of light. When a multi-frequency optical signal arrives at a Bragg Grating, the grating enbales all frequencies except the one for which it is "tuned" to pass through and reflects the frequency it is designed to select. The reflected signal is directed to another fiber. To create an optical switch, it is relatively straightforward to combine multiple Bragg Gratings in a series that is designed to filter different frequencies. By shunting the resulting frequencies in different directions, it becomes a simple matter to design an optical add-drop multiplexer, which is nothing more than a special purpose switch. It is only a matter of time before large-scale switches will use this technique, and all-optical devices will become commonplace in central offices.

SONET VERSUS DWDM: A TECHNOLOGY JIHAD. A battle is underway between those in the SONET camp and those in the DWDM camp over the underlying nature of the future transport network, and it is worthwhile to present both sides of the argument here. The argument centers on whether future networks will transport their payloads (probably IP packets) within the SONET framing structure or will simply hand them directly down to DWDM for transmission.

SONET is a physical layer protocol that defines a multiplexing scheme for the transport of a wide variety of services over a common optical network. It includes a well-defined network management system as well as a highly developed self-healing architecture to prevent catastrophic failures in the event of a span disruption. In the event that a ring is cut, the APS bytes kick in and cause the failed ring to be abandoned in favor of the backup ring, thus preventing an interruption of service.

DWDM, on the other hand, is a highly effective multiplexing scheme that is not yet widely standardized and that does not yet offer any sort of survivability guarantees along the lines of SONET's APS capability. The *Optical Internetworking Forum* (OIF) was created in April 1999 to develop a strategy for the widespread deployment of DWDM architectures and to guide the development of globally accepted standards that will ensure interoperability.

Today it is common to see the following architecture in high-speed networks. IP packets are carried within ATM cells, which are in turn transported within SONET frames over a DWDM-enhanced, dual counter-rotating, ring-based fiber network. This architecture is highly survivable and makes adding and dropping payload components extremely easy. It is also largely standards-based, because IP, ATM, and SONET are all based upon widely accepted international standards blessed by the *Internet Engineering Task Force* (IETF), the ATM Forum, and the ITU, to name a few. Meanwhile, the OIF has focused its energies on three critical areas for DWDM: circuit protection and restoration, integrated network management systems similar to those provided by SONET, and interoperability among optical interface equipment.

Although DWDM is perfectly capable of transporting IP packets at blindingly fast speeds, it does not yet have the capability to guarantee survivability, provide interoperability, or offer network management. And although ATM and SONET add a significant amount of overhead (and therefore inefficiency) to the transported signal, most believe that it is a fair price to pay for the peace of mind that comes from an overdesigned network that is not likely to fail. Because the majority of those looking to deploy DWDM-based fiber networks are the large, incumbent carriers whose primary focus has always been on customer service, they are not likely to surrender their ATM/SONET-based networks in the near term. However, with work underway to augment the DWDM technology with interoperability and survivability guarantees, it is possible that both SONET and ATM could become unnecessary in the future.

Some companies have already experimented with alternative architectures. In May 1999, Cambrian Systems combined their efforts with those of 3Com, Bay Networks, and Packet Engines to prove that SONET was not required for survivable transmission. Instead of ATM at the switching layer, they used gigabit Ethernet and created a highly efficient system that does not require the massive overhead that SONET brings with it. Widespread implementation, however, is a long way off. Traditional, widely accepted architectures will remain in use for some time to come. It was the telephone company, after all, that coined the phrase, "If it ain't broke, don't fix it." We will examine the SONET/DWDM debate in greater detail in a later chapter when we discuss QoS issues.

POINT-TO-POINT TECHNOLOGIES SUMMARY

Point-to-point technologies range from very low-speed, analog modem-based systems for telemetry and alarm circuit applications to unimaginably high bandwidth services that operate in the multiple gigabit-per-second range. For implementations that require the capability to transfer information between two non-changing points, or between a small number of points, point-to-point solutions represent a reasonable solution. They are secure, are universally available, offer minimal traffic delays, and are available with tremendous bandwidth granularity.

Point-to-point technologies have a number of downsides, however, that must be taken into account when selecting network architectures. First of all, they are dedicated, which means that once they are established they cannot be redirected to another site without a service order. Second, they are generally billed based on bandwidth and circuit miles, which means that the price does not change from month to month; there is no usage component. The circuit costs the same whether it is in use two percent of the time or 100 percent of the time, which means that it is most economical when it is in full-time use.

Although these downsides may appear trivial in light of the relative advantages of dedicated facilities, they can become major issues when flexibility, distance, and, in some cases, bandwidth concerns come into play. The result has been a migration to switched technologies that overcome all of these disadvantages while at the same time offering services that emulate private lines.

SWITCHED TECHNOLOGIES

Switched technologies came about because of the high cost and inflexibility characteristic of early dedicated communications architectures. When telephony was first introduced, there was no concept of switching. So few customers existed that it was common practice to run dedicated telephone circuits between individuals who wanted to call one another. As telephones became more common and call volumes increased, the need for switching arose. The first "switches" were people, usually women,[22] who asked the customer for the number they wanted to reach and then inserted a patch cord that interconnected the caller with the called party. In the early days, when call volumes were still relatively low, the technique of using human operators was adequate to handle the number of incoming call setup requests. However, as call volumes increased, human operators could no longer handle the task and mechanical switches were invented.

The first mechanical switch was invented in 1891 by Almon Strowger, a Kansas City undertaker. According to local legend, Strowger realized one day that his business volume had dropped off, and since the average death rate in Kansas City had not

[22]The first operators were actually young boys. They proved to be so rude and unmanageable that they were replaced with women operators whose interactions with customers proved to be far more professional.

declined, he became suspicious. Upon investigation he learned that the town's telephone operator was married to Strowger's competitor, which meant that when calls for an undertaker came in, they were not routed to Strowger. In response, he designed and built the first mechanical switch, called a *step-by-step*, which, when installed, gave customers the ability to make their own routing decisions by dialing numbers instead of depending on the operator to do it for them.

As time went on, mechanical step-by-step switches were replaced by various generations of faster, more efficient crossbar switches, which in turn were replaced by several succeeding generations of electronic switches. Modern central office switches such as the Lucent Technologies 5ESS, the Nortel Networks DMS100, the Siemens EWSD, and the Ericsson AXE are all examples of high-speed, high-volume devices capable of processing more than 100,000 calls per hour using a technique called *circuit switching*.

CIRCUIT SWITCHING

The telephone network is a perfect example of circuit switching. When a customer wants to place a call, they pick up the handset, which notifies the switch of their desire to do so, and the switch responds by sending a dial tone to the phone. The caller then sends the destination address to the switch (the telephone number), which proceeds to establish the call. As soon as the other end goes off-hook, the call is established.

The primary characteristic of this technology is its dedicated nature. For the duration of the call, all of the loop, switch, and trunk resources necessary to establish the call are dedicated to the two endpoints and cannot be shared with other users. As long as the call is in progress, the resources are off-limits to the network. As a consequence of this, circuit-switched networks must be massively overdesigned to accommodate the high volume of calls that they experience, particularly on holidays like Mother's Day. Circuit-switched networks are therefore expensive and inefficient, but extremely capable. The service they provide emulates the service delivered by a private-line circuit, albeit typically with less bandwidth than a high-end private-line circuit can provide. They work well for such applications as voice, low-speed video, medium-speed data and dial backup, and the technology is therefore deeply embedded in the network.

Circuit switching has experienced a major problem in recent years as Internet access with its attendant Web surfing has grown popular. Because call resources are dedicated, the use of circuit-switched facilities for modem access to the Web has resulted in the switches becoming severely overtaxed by the long call-hold times that characterize most Web sessions. When it becomes widely deployed, DSL will do a lot to eliminate the extra load on the switched network, but in the long run, packet switching represents the best solution for data transport. Circuit switching is still the best solution for voice traffic, particularly if the data packets can be routed elsewhere. Enter packet switching.

PACKET SWITCHING

Packet switching has its roots in a technique known as *store-and-forward switching,* which was first implemented as a technology called *message switching.* Store-and-forward systems do precisely what their name implies. Upon receiving information to be transported, the switch stores it in some form of memory, checks it for errors, makes a routing decision, and forwards the message out the proper port to the next switch.

Message switching is not used much anymore, but understanding how it worked helps to understand the mechanics of modern switch fabrics. In message switching, information to be transported across the network was sent from switch to switch as an entire message, a technique that has several downsides. First, it required the switches to have hard drives, because at the time messages were too large to store in the memory available in switches. This meant that the process was slow because of the need to involve a mechanical device to process the message. If a message arrived at a switch and upon examination was found to have errors, it was discarded and a second copy was requested from the prior switch. This is clearly wasteful; a single bit error could result in the need to retransmit an entire message. Early teletype systems relied on message switching, but the technology has largely been abandoned in favor of *packet switching*.

Packet switching is a data transmission technique in which messages (voice, video, images, or pure data) are broken into small segments called packets. Each packet is then encapsulated within a frame that may contain information about the nature of the data being transported, its relative level of criticality, its position within the data stream (one of five, two of five, three of five, and so on), some error correction information, and the addressing overhead.

Unlike their message-switched predecessors, packet-switched networks are quite efficient. They take into account the fact that data, as a general rule, is bursty, meaning that typical data users do not send continuous streams of information, but rather somewhat random bursts of traffic. As a consequence, they do not use the transmission facility 100 percent of the time; it may be idle for considerable periods of time. To overcome this inherent inefficiency, packet networks combine the packet streams from multiple users onto a single facility, thus using a higher percentage of the available bandwidth. Control protocols manage the independent data streams to ensure that each user data component maintains its integrity. In some cases, this technique is called *virtual circuit service*. The name stems from the fact that the service is so good in these networks that users feel as if they have their own dedicated transmission channel, in spite of the fact that the channel is shared among a collection of users. The word "virtual" becomes rather central in data communications, because it defines a family of quite lucrative services that don't really exist. Within the confines of communications, virtual has only one meaning: *It's a lie.* If you see the word used to describe a technology or service, you should immediately think, "They're lying to me. It isn't real." And yet, these services are extremely capable, cost-effective, and popular, as we will see.

CONNECTIONLESS PACKET-SWITCHED SERVICES. Packet networks provide two forms of service: connectionless service, where the switches do not perceive a relationship between packets that derive from the same source, and connection-oriented service, where they do. In connectionless networks, each packet contains complete destination address information and, as a result, every packet can be routed independently of the others. The benefit of this is that connectionless networks are extremely flexible since packets are not required to take any particular path through the network fabric and can be rerouted as the network sees fit. The disadvantage is that, because of this seemingly random routing mechanism, packets can arrive out of order, resulting in the need for a higher layer protocol to put them back into the correct order. Connectionless network protocols include IP among others, and the service they deliver is called *datagram.*

Connectionless networks make no guarantees of delivery or arrival sequence. They are often called "best effort" or "spray and pray" networks because, although they make every effort to deliver the packets to the destination, they do not guarantee the delivery. That is the responsibility of a higher layer protocol. For example, IP depends on TCP for iron-clad, guaranteed deliveries of transmitted packets if such guarantees are required. Their primary advantage comes from their inherent capability to route around trouble areas in the network, thus ensuring survivability.[23]

CONNECTION-ORIENTED PACKET-SWITCHED SERVICES. Connection-oriented networks differ from their connectionless counterparts in that they establish a path based on known network conditions and send all packets from the same source across the same path, thus ensuring sequential delivery. Because this service emulates private line, it is often called a *virtual circuit service.*

In connection-oriented networks, a multistage process must be adhered to before information is transmitted across the network.

---

[23]The Internet, which is connectionless, was originally conceived as a *Department of Defense* (DOD) communications network during the 1960s when the fear of nuclear attack was in the forefront of everyone's mind. IP was chosen because it would enable a network to continue operation even if one or more nodes were lost due to attack.

In stage one, a call setup packet (I like to call it a Lewis and Clark packet) is sent into the network. This initial packet contains the full address of the intended recipient of the message. Upon arrival at the first switch, the packet is examined, the address is read, and a route (outgoing port) is selected based upon known information about the network and the intended destination. The switch then makes an entry into its routing table, indicating which port the packet arrived on and through which one it went out. The packet continues across the network, causing table entries to be written at each switch, and thus blazes a trail across the network. Once the trail has been laid down, the remaining packets can make their way across the digital continent, following the blazes (table entries) left on the trees (switches) that they encounter along the way. As a result, the packets that follow the setup packet do not require a full address. All they need is a short virtual circuit identifier; the switches do the rest.

The advantage of connection-oriented packet switching is that, unlike connectionless service, all packets take the same route through the network, thus ensuring sequential delivery. The technology emulates the service a customer would receive from a dedicated facility. Examples of connection-oriented technologies include X.25 packet switching, frame relay, and ATM.

A word about packet technology: Once again, remember that this game is not about technology, but rather about providing customer service. The primary beneficiary of packet-switched technology is not the customer. If the service provider builds a robust, properly engineered packet-switched network that is capable of transporting a wide variety of service types throughout their operating area, and if they implement QoS protocols such that they can guarantee the services they sell to the customer, then they are in a position to be all things to all applications. If they then combine their advanced network fabric with a wide variety of access offerings, customers will have no reason to go anywhere else.

Remember the cloud? If service providers design the right cloud, they can satisfy every possible user demand for service, and customers will be happy to buy their "gozintas and gozoutas" from the service provider. If the customer wants to buy a private line service that operates at a particular speed, the service provider can deploy the circuit over a switched fabric that is *shared* but which gives the appearance of being *dedicated*. By taking advantage of the inherently efficient nature of packet technology, the service provider can build a far more cost-effective network while continuing to satisfy customer demands for high-quality, dependable service. Ultimately, the customer doesn't care and doesn't have to know. All they care about is that they are getting what they asked for, and the underlying technology, while critically important, remains invisible to them.

In order for this concept to work, a newer form of packet technology must be introduced. Datagram and traditional virtual circuit services are fine for the limited requirements of low and medium-speed data—the services they were originally designed for—but they fall short of being able to provide for modern applications that require greater bandwidth allocations. For those services, *fast packet technology* is required.

FAST PACKET SERVICES.   Fast packet technologies include such services as frame relay and ATM and are considered fast for several reasons. The principle reason is that they have the capability to dramatically reduce the overhead required to receive, examine, and forward a packet of data by as much as 80 percent.

In order to achieve these remarkable improvements, fast packet technologies make a number of assumptions about the network. First, they assume that it is based upon a fiber infrastructure from which they can further assume a low error rate. Second, they assume modern, robust switch fabrics that also offer low error rates. Finally, they make the assumption that the end devices possess a modicum of intelligence and can thus make informed decisions about the "correctness" of the received data.

When a traditional packet switch receives a packet, it goes through the following steps. First, it buffers the packet. Next, it invokes an error detection algorithm and checks the packet for bit errors. If it finds one, it invokes an error *correction* algorithm, which in most cases causes the switch to go back to the last switch in the chain and ask for another copy of the packet while at the same time discarding the bad one. Once it receives a copy that it determines to be error-free, the switch goes to its routing table, selects a route for the packet, and transmits it to the next switch in the chain.

In fast packet environments, the process is dramatically simplified. The switch receives a packet, and rather than buffer the packet, it often implements an option called *cut-through*. With cut-through invoked, the switch never buffers the packet but rather proceeds to parse it for errors while at the same time examining the destination address information and selecting an outgoing port. In effect, the head of a packet can already be on its way out of the switch while the tail is still being examined for errors. In the event that an error is detected, there is no attempt to correct it. In fact, the switch does not ask for a resend, nor does it attempt to notify anyone of the bad packet. It simply throws the bad packet away and makes the correct assumption that the end stations have enough innate intelligence to detect the error and take whatever steps are necessary to correct it. This process is significantly more efficient than the store-and-forward technique described earlier.

Of course, in reality there are no error-free networks—it just isn't possible. But the error rate in fast packet networks is low enough that the inefficiency that results when an error *does* occur is overcome by the benefits of the streamlined network. As we mentioned earlier, frame relay and ATM are the most common examples of fast packet switching.

FRAME RELAY. Often described as X.25 on steroids, frame relay came about as a private-line replacement technology and was intended as a data-only service. Today it routinely carries not only data, but voice and video as well, and although it originally emerged with a top speed of T-1/E-1, it now provides connectivity at much higher speeds. In frame relay networks, the incoming data stream is packaged as a series of variable length frames[24] that are the telecommunications equivalent of a Trojan Horse. A frame relay frame can transport any kind of data: LAN

---

[24]In telecommunications parlance, a frame is a distinguishable entity that has a clear beginning and end, a variable length field for application data, and some form of error detection mechanism. In many cases, the user data contained within a frame is a packet.

traffic, IP packets, SNA frames, even voice and video in certain cases. In fact, of late it has been recognized as a highly effective transport mechanism for voice, enabling frame relay-capable PBXs to be connected to a frame relay network, which can cost-effectively replace private-line circuits used for the same purpose. When voice is carried over frame relay, it is usually compressed and packaged in small frames for transport to minimize the processing delay of the frames. According to the Frame Relay Forum, as many as 255 voice channels can be encoded over a single PVC, although the number is usually smaller when implemented.

Frame relay is a virtual circuit service. When a customer wants to connect two locations using frame relay, they call the service provider and indicate to the service representative the locations of the endpoints and the bandwidth they require for the applications that will operate across the circuit. The service provider issues a service order that results in the establishment of the circuit. If, at some point in the future, the customer decides to change the circuit endpoints or upgrade the bandwidth, another service order must be issued. This service is called *Permanent Virtual Circuit* (PVC) and is the most commonly deployed frame relay offering.

Frame relay is also capable of supporting a *Switched Virtual Circuit* (SVC) service, but SVCs are not currently available. With an SVC service, customers can make their own modifications to the circuit by accessing the switch and requesting changes. However, service provides do not currently offer the service because of billing and tracking issues. Instead, they allow customers to create a fully meshed network between all their locations that allows any user on the network to send traffic to any other user on the network. Instead of making routing changes in the switch, the customer has a circuit between every possible combination of desired endpoints. The service provider usually makes this a cost-effective option for the user by charging a normal monthly fee for the first PVC and low fees for each additional PVC. As a result, customers get the functionality of a switched network, while the service provider avoids the difficulty of administering a network within which the customer is actively making changes.

In frame relay, PVCs are identified using an address called a *Data Link Connection Identifier,* (DLCI, pronounced "del-sie"). At any given endpoint, the customer's router can support multiple DLCIs, and each DLCI can be assigned varying bandwidths based upon the requirements of the device/application on the router port associated with that DLCI.

In Figure 1-19, the customer has purchased a 768-Kbps circuit to connect their router to the frame relay network. The router is connected to a videoconferencing unit at 384 Kbps, a frame relay-capable PBX at 384 Kbps, and a data circuit for Internet access at 128 Kbps. Note that the aggregate bandwidth assigned to these devices exceeds the actual bandwidth of the access line by 128 Kbps. Under normal circumstances, this would not be possible, but frame relay assumes that the traffic that it will normally be transporting is bursty by nature. If the assumption is correct (and it usually is), there is little likelihood

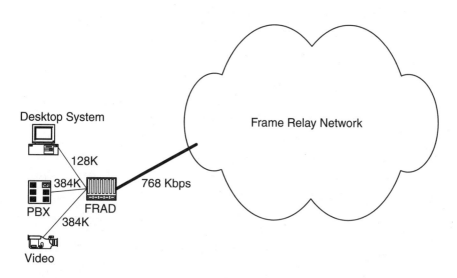

FIGURE 1-19 A typical frame relay network

that all three devices will burst at the same time. As a consequence, the circuit's operating capacity can actually be over-
booked, a process known as *oversubscription*. Most service
providers allow as much as 200 percent oversubscription, something customers clearly benefit from, provided the circuit is
designed properly. This means that the salesperson must carefully assess the nature of the traffic that the customer will be
sending over the link and ensure that enough bandwidth is allocated to support the requirements of the various devices that
will be sharing access to the link. Failure to do so can result in
an under-engineered facility that will not meet the customer's
throughput requirements. This is a critical component of the
service delivery formula.

The throughput level, that is, the bandwidth that frame
relay service providers absolutely guarantee on a PVC-by-PVC
basis, is called the *Committed Information Rate* (CIR). In addition to CIR, service providers will often support an *Excess
Information Rate* (EIR), which is the rate above the CIR they
will attempt to carry, assuming the capacity is available within
the network. However, all frames above the CIR are marked as
eligible for discard, which simply means that the network will
do its best to deliver them but makes no guarantees. If push
comes to shove, and the network finds itself to be congested,
the frames marked *discard eligible* (DE) are immediately discarded at their point of ingress. This CIR/EIR relationship is
poorly understood by many customers because the CIR is taken
to be an indicator of the absolute bandwidth of the circuit.
Whereas bandwidth is typically measured in bits per second,
CIR is a measure of *bits in one second*. In other words, the CIR
is a measure of the average throughput that the network will
guarantee. The actual transmission volume of a given CIR may
be higher or lower than the CIR at any point in time because of
the bursty nature of the data being sent, but in aggregate, the
network will maintain an average guaranteed flow volume for
each PVC. This is a selling point for frame relay. In most cases,
customers get more than they actually pay for, and as long as
the switch is properly engineered, it will not suffer adversely

from this charitable bandwidth allocation philosophy. The key to success when selling frame relay is a clear understanding of the applications the customer intends to use across the link so that the facility can be properly sized for the anticipated traffic load.

SERVICE GRANULARITY IN FRAME RELAY.   Frame relay does not offer a great deal of granularity when it comes to QoS. The only inherent mechanism is the DE bit, described earlier as a way to control network congestion. However, the DE bit is binary; it has two possible values. Thus, a customer has two choices: the information being sent is either important or it isn't—not particularly useful for establishing a variety of QoS levels. Consequently, a number of vendors have implemented proprietary solutions for QoS management. Within their routers (sometimes called *Frame Relay Access Devices* [FRADs]) they have established queuing mechanisms that enable customers to create multiple priority levels for differing traffic flows. For example, voice and video, which don't tolerate delay well, could be assigned to a higher priority queue than one to which asynchronous data traffic would be assigned. This allows frame relay to provide highly granular service. The downside is that this approach is proprietary, which means that the same vendor's equipment must be used on both ends of the circuit. Given the strong move toward interoperability, this is not an ideal solution, because it locks the customer into a single vendor situation.

CONGESTION CONTROL IN FRAME RELAY.   Frame relay has two congestion control mechanisms that should be mentioned. Embedded in the header of each frame relay frame are two additional bits called the *Forward Explicit Congestion Notification bit* (FECN) and the *Backward Explicit Congestion Notification bit* (BECN). Both are used to notify devices in the network of congestion situations that could affect throughput.

Consider the following scenario. A frame relay frame arrives at the second of three switches along the path to its intended destination where it encounters severe local congestion. The congested switch sets the FECN bit to indicate the presence of congestion and transmits the frame to the next switch in the chain. When the frame arrives, the receiving switch takes note of the FECN bit, which tells it the following: "I just came from that switch back there, and it's extremely congested. You can transmit stuff back there if you want to, but there's a good chance that anything you send will be discarded, so you might want to wait awhile before transmitting." In other words, the switch has been notified of a congestion condition that it may respond to by throttling back its output to give the affected switch time to recover.

On the other hand, the BECN bit is used to control a device that is sending too much information into the network. Consider a situation where a particular device on the network is transmitting at a high volume level, routinely violating the CIR and perhaps the EIR level established by mutual consent. The ingress switch, that is, the first switch the traffic touches, has the capability to set the BECN bit on frames going toward the offending device, which carries the implicit message, "Cut it out or I'm going to hurt you." In effect, the BECN bit notifies the offending device that it is violating protocol and continuing to do so will result in every frame from that device being discarded without warning or notification. If this happens, it gives the ingress switch the opportunity to recover. However, it doesn't fix the problem; it merely forestalls the inevitable, because sooner or later the intended recipient will realize that frames are missing and will initiate recovery procedures, which will cause resends to occur. However, it may give the affected devices time to recover before the onslaught begins anew.

Ultimately, FECN was designed for destination-controlled flow control schemes, such as TCP; BECN was intended for use by source-controlled flow control schemes such as SNA.

The problem with FECN and BECN is that many devices choose not to implement them, choosing instead to rely on the Sylvester Stallone protocol. Allow me to explain. Many devices, upon receiving a frame with a set FECN or BECN bit, respond with the digital equivalent of (Sly's voice, now), "Yeah. So?" They do not necessarily have the inherent capability to throttle back upon receipt of a congestion indicator, although devices that can are becoming more common. Nevertheless, proprietary solutions are in widespread use and will continue to be so for some time to come.

FRAME RELAY SUMMARY.   Frame relay is clearly a Cinderella technology, evolving quickly from a data-only transport scheme to a multiservice technology with diverse capabilities. For data and some voice and video applications, it shines as a WAN offering. In some areas, however, frame relay is lacking. Its bandwidth is limited to DS-3 and its capability to offer standards-based QoS is limited.

ASYNCHRONOUS TRANSFER MODE (ATM).   Like frame relay, ATM is a connection-oriented technology that has assumed the mantle of leadership among current WAN offerings for a number of reasons. Frame relay and ATM share both similarities and differences. Both are virtual circuit-based fast packet services for wide area transport that support a variety of payload types. Whereas frame relay's bandwidth is limited to 45 Mbps, ATM tops out at the upper limits of SONET speed, currently OC-192 (approximately 10 Gbps). Where frame relay requires proprietary solutions for granular QoS, ATM offers an array of internationally accepted, widely standardized QoS definitions that enable it to support a plethora of services with ease, in spite of being characteristically different in terms of relative levels of necessary bandwidth, burstiness exhibited, and QoS required. Furthermore, unlike frame relay, ATM's "transmission unit" is a fixed-size cell rather than a variable-size frame. This has a number of advantages. First, it simplifies the design of the switch. In frame relay, the switch must be capable of processing frames that arrive in a variety of sizes, which makes buffering and switching a complex process. In cell relay, of which ATM is an example, all cells are exactly the same size (a five-octet header that contains information the switch uses to properly route the cell, and a 48-octet payload field that contains the user's data). Regardless of the payload type, the switch can always expect to receive protocol data units of the same size, which means that its buffering is simplified and the switch is therefore very fast.

ADDRESSING IN ATM.   In ATM, virtual circuits are identified by a combination of *Virtual Path Identifiers* (VPI) and *Virtual Channel Identifiers* (VCI). A single virtual circuit is identified by a VPI/VCI combination. VCIs are unidirectional channels, while a VPI is a combination of VCIs that results in a bidirectional circuit. When a device wants to send traffic to another device in an ATM network, it addresses the message to the appropriate virtual path that by definition identifies the virtual channels.

QOS AND COS CONTROL IN ATM.   The terms QoS and *Class of Service* (CoS) are used interchangeably in discussions about transport technologies. They are, however, quite different, and it would be useful to make the distinction here.

QoS is a measure that is inviolable. It is absolute in the sense that when QoS is implemented in a network the provider guarantees that they will deliver measurable performance characteristics as defined by the QoS statements in the *Service Level Agreement* (SLA) that exists between the provider and the consumer of the service.

CoS, on the other hand, is defined by the provider of the service and is therefore a relative measure, unrelated to measures of quality. For example, a service provider would be perfectly justified (although not particularly bright) if they were to identify and advertise the following transmission CoSs:

- Really good
- Pretty good
- Not so great
- Fair
- So-so
- Better write a letter

More reasonable classes might define services that align accurately with the types of application traffic transported by the network provider for the customer.

ATM has experienced several generations of service class definitions over the years, including efforts by both the ITU and the ATM Forum. By and large, all the efforts have revolved around three service characteristics: whether the traffic's bit rate is constant or variable, what kind of timing relationship exists between the end points of the facility, and whether the circuit is connectionless or connection-oriented. Using these three characteristics, the ITU developed the model shown in Figure 1-20. Known as the *Broadband ISDN Protocol Model*, it enables service providers to define classes of service that meet the transport requirements of every possible traffic type. Broadband ISDN has little to do with traditional narrowband ISDN, other than the fact that it defines a broadband digital network over which a set of integrated services can be transmitted.

A	B	C	D
Connection-Oriented			Connectionless
Constant	Variable		
Required		Not Required	
Circuit Emulation (Voice)	Packet Video	Connection-Oriented Data (Frame Relay)	Connectionless Data (IP)
1	2	3	4

FIGURE 1-20 The original ITU ATM service model. Note that AAL 5 is not shown; it will be introduced later.

Over time, and as a result of a shared effort between both the ITU and the ATM Forum, the parameters for assigning service definitions (QoS) evolved. Similar to frame relay, ATM now relies on cell rate indicators to identify service flows within the network.

To understand QoS in ATM, it is useful to compare it to frame relay. Frame relay relies on a CIR to indicate the guaranteed bandwidth of each PVC. Instead of a CIR, ATM relies on a *Sustainable Cell Rate* (SCR), which defines the guaranteed cell rate that a station can transmit into the network. On the other hand, the *Peak Cell Rate* (PCR) is analogous to frame relay's *Excess Information Rate* (EIR), defining the maximum traffic rate acceptable by the network. Finally, the maximum burst size defines the maximum cell volume acceptable within a defined period of time, usually one second.

Based on the characteristics described previously, ATM defines a series of service classes that accommodate the requirements of all application types. *Constant Bit Rate Service* (CBR) defines a class of service that emulates the constant delay, high-priority nature of a dedicated, private-line circuit. *Variable Bit Rate Service* (VBR) guarantees that traffic falling within the SCR will be carried, while anything above the SCR will be marked DE and may be discarded if it exceeds the maximum burst size.

Variable Bit Rate-Real Time Service (VBR-rt) is similar to VBR but adds better congestion control and latency guarantees. It is typically used for the transport of voice and video, which require minimal delay through the network.

Available Bit Rate (ABR) is a variation on a theme. Although it does not make any guarantees relative to delay, it does guarantee bandwidth availability. In ABR, cells may be buffered by the network, and in some cases, the network may attempt to control an offending device. As a result, ABR is something of a hybrid of CBR's guaranteed transport and VBR's flexibility.

ABR also contains a bit of the final service class, *Unspecified Bit Rate Service* (UBR). UBR makes no guarantees at all, similar to IP. In spite of what may appear to be a less-than-desirable service choice, UBR is widely deployed.

CONGESTION CONTROL IN *ATM*.   Flow management in ATM has been widely studied, and two principal mechanisms based on standards have resulted. The first is the use of the *Explicit Forward Congestion Indication* (EFCI) *bit*, found in the ATM cell header, which is used to notify downstream switches of congestion. When a switch receives a cell that has the EFCI bit turned on, the receiving switch responds by sending a *choke cell* to the transmitter (known as a *resource management* [RM] cell) that attempts to slow the transmission volume. The alternative flow control technique is called *explicit rate,* a more elegant scheme than the rather primitive EFCI solution. When explicit rate is implemented, a resource management cell is inserted in the cell stream after every 32nd cell. These indicator cells enable receiving ATM network components to respond more quickly, because they can detect a pattern in the information contained in the RM cells that might indicate a congestion buildup.

Attentive readers may have noticed that although ATM has an EFCI, there is no EBCI. The reason for this is that congestion control is done on a virtual channel basis, and because virtual circuits are unidirectional, there is no need for backward congestion notification.

So, WHY ATM?   If we were to design the perfect network, it would

- Operate at a range of speeds
- Exhibit a low error rate over long operational periods
- Handle multiservice and multiprotocol support
- Be based on international standards
- Support bandwidth on demand
- Be imminently scalable
- Manage traffic well during periods of congestion or other disruptive events
- Be interoperable with other technologies
- Offer diverse QoS
- Be "future-proof"
- Have a well-developed network management scheme

Let's consider ATM in comparison with the list of capabilities shown previously.

Operate at a range of speeds. ATM is part of the broadband ISDN protocol model, designed to operate at SONET/SDH speeds. This includes bandwidth provisioned from OC-1 (51.84 Mbps) to OC-192 (9.95 Gbps). In the unlikely event that more bandwidth is needed, DWDM can provide multiples of SONET speeds.

Exhibit a low error rate over long operational periods. ATM is a fast packet technology, which, by definition, experiences extremely low incidences of errors. Furthermore, it interfaces directly with SONET, which, as an optical technology, is equally free of the typical errors that plague metallic facilities.

Handle multiservice and multiprotocol support. ATM's well-defined service classes (CBR, VBR, ABR, and UBR), as well as the diverse service and parameter qualifiers that the protocol stack relies on (timing sensitivity, bit rate variability, connection mode, sensitivity to burstiness, and so on) give it the unique capability to not only transport a wide variety of service types, but to guarantee that the delivery quality they expect based on the nature of the application they derive from is assured.

Be based on international standards. Both the ITU and the ATM Forum have contributed to the development of standards for ATM that are recognized worldwide.

Support bandwidth on demand. The well-developed signaling scheme that underlies the ATM protocol model enables the allocation of bandwidth on an as-needed basis. Because SONET is often the assumed physical layer, there is never a question about the availability of bandwidth when an application requires it.

Be imminently scalable. Because ATM appears to assume the role of a primary core technology for the next-generation network, manufacturers of ATM switching equipment have made scalability a key tenet of their design efforts. Like central office switches, ATM switches can be augmented to respond to customer growth as required.

Manage traffic well during periods of congestion or other disruptive events. Because ATM is much more than an asynchronous data transport technology, it must be able to handle the mixed performance requirements of both delay-sensitive and -insensitive services. Consequently, ATM supports a variety of network management protocols that have the capability to perform flow control and to route around problem areas in the network should the need arise.

Be interoperable with other technologies. ATM is not particularly discriminating as technologies go and is, in fact, rather forgiving. It has the capability to "emulate" other technologies and can thus provide transport between them when required. For example, ATM offers a popular service called the *Frame Relay Bearer Service* (FRBS). With FRBS, service providers can deploy an ATM network from which they can provision both ATM and frame relay services. This is accomplished with ease. If a customer wants to buy frame relay service from the provider, the provider simply adds frame relay interface cards to the ATM switch (see Figure 1-21). The customer sends frames to the switch, which converts them to cells for long-haul transportation. At the other end of the circuit, the ATM switch converts the cells back to frames for delivery to the intended recipient. In this situation, the customer has no idea that their traffic is actually being transported by a far more efficient network fabric than ATM. The good news is that they don't *have* to know. From a service's point of view, it doesn't matter.

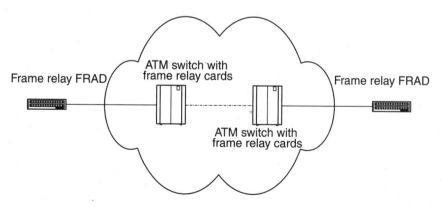

FIGURE 1-21   Frame Relay Bearer Services (FRBS)

Offer diverse QoS. The primary reason that ATM has emerged as the recommended foundation technology for the deployment of IP networks is its capability to offer service granularity. No one will argue with the observation that ATM is *not* an ideal transport choice for voice, video, medical images, or data. Clearly, circuit switching offers a better solution for voice, while a private line circuit is much more appropriate for the rigorous demands of video or medical imaging files. ATM, however, provides the ideal solution for the broad mix of all those services. However, if a service provider wanted to deploy a single-fabric network that could satisfy the bulk of its customers' needs while taking into account cost-effectiveness, ATM would be the only contender.

Be "future-proof." ATM is. No technology looming on the immediate horizon threatens ATM's supremacy. Many believe that IP will ultimately be retrofitted with the capability to provide diverse QoS, but until that happens ATM is the only answer that makes complete sense.

Have a well-developed network management scheme. ATM network management offers well-designed maintenance, monitoring, and provisioning capabilities. When combined with the inherent capabilities of SONET, the result is a well-run network.

THE ATM MARKET.   For a long time, ATM's position in the marketplace was sketchy because it was considered an expensive alternative, especially in the local area. Over time, however, its proper position has become clear. ATM provides dependable, high-volume bandwidth for large-scale enterprise networks. In smaller corporate networks, it provides an aggregation solution for lower bandwidth technologies such as T-1 and DS-3. Today the cost for ATM services has dropped to the point that it is approaching parity with frame relay. Most interexchange carriers offer ATM service at prices comparable to frame relay, and in some cases cheaper. IXCs by and large charge slightly more for ATM but expect to offer the two at similar prices soon.

Both PVC and SVC services are available from ATM, but PVC is far more widely deployed. As in frame relay, SVCs are difficult to measure, track, and bill, and are therefore eschewed by most carriers. Some do offer the service, however. Sprint and Qwest bill their SVC customers according to cell volume, while AT&T bills based on connection duration.

ALTERNATIVES TO ATM.   We have already mentioned the fact 3tocol. That is not likely to happen in the near term, but two alternatives have emerged that could prove to be viable challengers.

Cisco has released its *Spatial Reuse Protocol* (SRP) as well as devices that implement it. With SRP, routers can be interconnected by dual-fiber rings using SONET and/or DWDM. The rings offer redundancy and operate at 10 Gbps with a switchover speed of less than 50 ms. In MAN scenarios, SRP could dramatically lower the cost of network deployment for corporations or ISPs looking to interconnect multiple sites by eliminating the need for an expensive ATM infrastructure. The one thing SRP does not adequately offer today, however, is a well-defined strategy for QoS, something ATM does extremely well.

The second alternative is called *Dynamic Synchronous Transfer Mode* (DTM). Introduced by three Swedish firms in January 1999, DTM is similar to circuit switching in that it relies on timeslots, but is different in that it can reallocate those timeslots on an as-needed basis to meet the demands of bursty applications when they arise.

ATM SUMMARY.   ATM has established itself as a reliable and flexible core switching technology for the vast majority of service types. It relies on international standards, provides good QoS and CoS control, has a robust embedded network management scheme, and is future-proof enough to handle the growing demands of new applications for some time to come. What it lacks, however, is a widely accepted and flexible addressing scheme and the universality that other protocols enjoy. It is an expensive solution when deployed at the edge of the network, but it is quite cost-effective when used as a core WAN technology. Ideally, ATM, when combined with another protocol that overcomes its few shortcomings, would be an unbeatable combination.

THE INTERNET PROTOCOL (IP)

IP is part of the well-known TCP/IP protocol suite that came about in the 1960s as an integral part of the Department of Defense's *Advanced Research Projects Agency Network* (ARPANET) project. It is a network layer protocol, responsible for routing functions and congestion control, and works closely with its transport layer counterpart, the *Transmission Control Protocol* (TCP). TCP/IP supports a wide array of application services and will operate over many different physical media types.

IP is a connectionless network layer protocol, which means that it does not perform a call setup prior to transmitting data packets. Instead, it transmits the packets into the network, each bearing a complete destination address, and trusts the network to deliver them to their final destination. With the help of various routing protocols, the packets do generally arrive, but because the service is connectionless and because connectionless networks do not discern a relationship between packets that originate from the same source, they may take different routes from source to destination based on changing network congestion and other factors. Consequently, they may arrive out of sequence. The advantage of a connectionless network, of course, is that it has the capability to avoid troubled areas in the network by dynamically routing around them. This flexibility is worth the cost of connectionless service, particularly because the transport layer (TCP) will generally correct any discrepancies that arise from IP's connectionless nature.

Addressing in IP

The current version of IP, known as *IP Version 4* (IPv4), relies on what is known as a *dotted decimal addressing scheme*. The name derives from the fact that IP addresses comprise four eight-bit segments separated by periods, as shown:

255.255.255.255

Each IP address is 32 bits long (four eight-bit components), which means that 2^{32} (roughly 4.2 billion) possible addresses can be created. The good news is that 4.2 billion is *a lot*. The bad news is that it isn't enough! Because of the immense and unexpected popularity of IP as a universal addressing scheme, we are getting dangerously close to exhausting the available address space. There are a number of reasons for this. The diversity of devices that can actually be addressed by IP is growing, networked devices in general are growing, and the manner in which IP addresses are assigned is not particularly efficient.

To understand the challenges that face IP addressing, it is first necessary to understand how the addresses work. Five classes of IP address exist, labeled A, B, C, D, and E. Classes A through C are used by network hosts, while Class D is reserved for multicasting and E for future use. They are created as follows. In any given network, it is necessary to identify both the network that serves that company as well as the devices attached to that network. By assigning the four eight-bit pieces of the IP address in a variety of ways, we can create a tiered address scheme that enables us to identify a small number of addresses with many hosts (devices), a large number of networks with a few hosts, or anything in between.

Consider Figure 1-22. In a Class A address, three of the bytes are assigned as host identifiers, while one is used to identify networks. Consequently, a Class A address can identify as many as 16 million unique hosts (over 2^{24}) and up to 126 (2^7-2) unique networks. As it turns out, not all the bits in each byte are actually used for addressing. Some are reserved to indicate the class of the address. For example, in Class A, the most significant bit is set to 0 to indicate that it is, in fact, a Class A address, leaving seven bits to identify the network. Some addresses are reserved, so the maximum number of identifiable networks in a Class A address is 126. Similarly, a Class B address can identify 65,534 hosts on 16,382 networks, while a Class C address, the most commonly deployed of all, can identify 254 hosts on more than 16 million networks.

The biggest problem facing IP today is the fact that when addresses are assigned, an entire address block is issued based on the number of hosts on a network. For example, a network of 50 hosts would be granted an entire Class C address, which would allow them to easily accommodate their 50 employees, but would

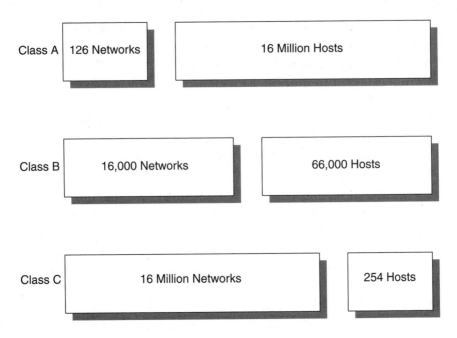

Class A | 126 Networks | 16 Million Hosts

Class B | 16,000 Networks | 66,000 Hosts

Class C | 16 Million Networks | 254 Hosts

FIGURE 1-22   IP address structure

leave a significant number of IP addresses within the range idle. This is a serious problem: some reports estimate that as few as 20 percent of all assigned IP addresses are actually in use.

To combat this problem, a number of solutions have been devised. The first we will describe is called *subnet masking*. In subnet masking, IP addresses are broken into three pieces instead of two (NETID and HOSTID). The first piece identifies the network; the second, a subnet address; and the third, the host on that subnet. Each subnet is limited to 254 nodes, but within a Class B address, for example, there can be as many as 256 uniquely identifiable subnets. In effect, the network is sub-divided into pieces, making management and administration easier, and far more granular. By creating a larger number of unique networks within the address space normally reserved for one network, better assignment of IP addresses is possible.

The second technique that is being used is a protocol called *Dynamic Host Configuration Protocol* (DHCP). DHCP enables a server to dynamically assign IP addresses to requesting stations, thus eliminating the problem of unused IP address space.

The third solution is called *Classless InterDomain Routing* (CIDR) and Variable Length Subnet Masks (VLSM). CIDR (pronounced like the drink) enables Class C addresses to be broken into smaller pieces than 254 hosts. This helps to resolve the problem of IP address overassignment. If a network has 50 hosts, they can be allocated a block of 64 adresses, for example, rather than an entire Class C block, thus making the address assignment and usage process far more efficient.

Although originally designed for address consolidation and as a way to simplify routing tables, VLSMs can also be used to allocate these sub-Class C address blocks.

Finally, the solution that many are waiting for is the next version of IP, called *IP Version 6* (IPv6).[25] Sometimes called IPng (for *next generation*), IPv6 adds enhancements to IPv4, including 128-bit addressing, dynamic reconfiguration, enhanced network security, multicasting, better QoS discrimination, and a number of other features. With 128 bits of addressing space, 2^{128} unique addresses can be created, which is approximately 10^{38} in total. One difficulty that has been identified is the conversion from IPv4 to IPv6. Many experts believe that the conversion is a necessary evil that must be faced, and that the deployment of *Network Address Translation devices* (NATs) will be required for some time. Indications from the first implementers of IPv6 are that the advantages far outweigh the downsides of the conversion.

WHY IP?

Of all the questions answered in this book, "Why IP?" is most pressing today. IP has enjoyed more press and attention than any other technology topic in the last few years, and for good reason.

One of the biggest problems that network services providers have traditionally faced is the issue of divergence. They offer a

[25]IPv6 is described in IETF RFC 1752.

wide range of services including POTS, ISDN, X.25 packet transport, frame relay, ATM, high-speed point-to-point services based on SONET, and many others. They offer them because customer applications demand them.

The problem with having such a diverse product set is that it requires an equally diverse support infrastructure. Each of these technology-based services requires billing, provisioning, installation, administration, maintenance, troubleshooting, repair, and direct customer contact support. This divergence is expensive, cumbersome, time-consuming, and *not* in the best interest of either the customer or the service provider.

Imagine a scenario, then, where a service provider could deploy a single network fabric from which they could provide all of their offered services, guarantee high quality, enjoy lower costs in every facet of the business, and easily meet the evolving needs of all customers. In 1982, that is precisely the role that the service provider (all one of them) played. If a customer came to AT&T with a business problem in search of a telecommunications solution, AT&T would quite comfortably and reliably respond, "Look, you just tell us where the terminals are, where the host is, and where the bank branches are, and we'll take care of the stuff in the middle. That's our job." As a service provider, they did not burden the customer with cloud-related stuff, because they *knew* that their network could reliably meet the demands of the customer, regardless of the nature of those demands.[26] The customer had no reason to go anywhere else, because they knew that AT&T would deliver a solution that worked.

The fact of the matter is that service providers need to get back to that 1982 model. OK, not the monopoly model, but a model that places them in the position of being a service provider that is so good at providing service that the customer wouldn't think of going anywhere else. One way to do that is to

---

[26] I am not looking to start a debate here on the merits or evils of monopoly business. Just consider the service implications of what I am describing.

build a communications delivery infrastructure that converts *divergence* to *convergence*. If service providers can deploy a technology base that dramatically simplifies the manner in which they deliver services, then untold advantages result, including those listed earlier such as lowered costs, shorter time to market, and diverse services.

MAKING IT HAPPEN

Let's go back and revisit our perfect network model once again. As we demonstrated, ATM, as a lower-level switching fabric, uniquely satisfies all of a perfect model's requirements, but at least three other characteristics exist that ATM does not provide that must be considered. All of them derive from a higher-layer network protocol.

First, the network protocol employed must be widely understood, universally available, inexpensive to implement, and simple to manage and administer. Second, it must offer an addressing scheme that will support large and growing population bases and address a wide variety of device types. Third, it must be universally application-friendly.

IP meets these requirements quite effectively. It is one of the most widely deployed protocols on Earth, found in LANs, MANs, and WANs, as well as in every modern network operating system. In terms of ubiquity, IP has no equal. Furthermore, because of its universal acceptance, it is widely understood, and sophisticated tools exist to manage networks based on it.

As an addressing scheme, IPv4 certainly has its weaknesses. However, they are rapidly disappearing under the onslaught of such solutions as NAT, CIDR, subnet masking, and, of course, the inevitable arrival of IPv6. The result of all this is that IP has become the only universal addressing scheme we have. It is accepted worldwide, has more than enough capacity, and, thanks to efforts by a plethora of vendors, is now interoperable with network operating systems, signaling systems, and most applications that require it.

IP/ATM INTEGRATION

So what's the conclusion we're looking for? If a service provider wants to deploy a full-service network, a likely solution may comprise an ATM core with IP running on top of it. By combining the advantages of ATM with those of IP, service providers can create a universal cloud that really does it all from end-to-end. IP brings the statistical loading advantages of a packet network as well as addressing and universality, while ATM adds QoS, bandwidth, and standards-based technology in the transport core. With this combination of technologies, network providers can make every possible service available including dedicated private lines, frame relay, ATM, voice capabilities, asynchronous data transport, interactive and distributive video, and many others. The elegance of this solution is that the underlying technology need not concern the customer, because ideally the service provider no longer sells technology. They sell service, which relies, of course, on technology. The customer, however, is not burdened with the details.

Consider the following scenario. Some day in the not-too-distant future, XYZ ILEC decides that they are technologically ready to convert to an all-IP network. They have relied on ATM for a long time to provide the solid QoS support needed to satisfy customer demands, but now new QoS protocols (discussed in the next section) have advanced to the point that ATM is no longer necessary in every case. Those QoS protocols, combined with IP and a high-speed, robust physical infrastructure, obviate the need for ATM's considerable capabilities. So, late one night, at a pre-appointed time, XYZ's central office technicians take a deep breath and pull the switch, converting the company to an all-IP infrastructure. And the best part of all this? *The customer never knows it has happened.* Services go in; services come out. The nature of the underlying fabric is completely invisible to the customer, as it should be. Service quality does not change from a customer's point of view, but from the service provider's perspective, life just became remarkably more simple. A single network now does it all.

The evolution to bring this about is already underway, and major vendors are lining up to meet the demand and become the center of the universal services cloud. Lucent Technologies, for example, has announced the 7 R/E, the successor to its venerable flagship the 5ESS. The design of the 7 R/E recognizes the enormous embedded value of the 5E platform, but also recognizes that circuit switching is no longer the answer to every problem. Built around an IP/ATM core, the switch combines the stability and latency-free service of circuit switching with the robustness, flexibility, and service transparency of IP. ATM serves as the core transport mechanism, but the time may come when ATM is no longer required, at which point IP may take over. The 7 R/E is fully capable of allowing that to happen when the time comes.

As the name implies,[27] the 7 R/E can be used as an immediate full-service switching solution for green field networks[28] or as an evolutionary solution for legacy installations. In general, the 7 R/E enables the embedded 5E base to continue with its evolution because, in spite of claims to the contrary, circuit switching is still a viable, growing solution for voice applications. As the network continues to advance, the 5E can "evolve" to a 7 R/E as required, but the design of the switch enables the evolution to be graceful, modular, and timely. This lets service providers maximize the utility of their investment in circuit switching while at the same time moving forward with packet-based IP. Finally, an integral network management system completes the package.

Nortel Networks has a similar plan underway with its Succession™ product line. The platform facilitates the convergence of voice and data, transforming the traditional circuit-switched network into a distributed broadband network that delivers seamless services over an IP/ATM backbone optimized

[27]The R stands for Revolution, the E for Evolution.

[28]A green field market is one that is new and does not yet have an embedded infrastructure.

for the transport of voice traffic, thus ensuring the protection of highly lucrative legacy services while at the same time creating a viable platform for the delivery of broadband. The switch is based on international standards and offers many advantages, including compatibility with multiple vendors, reduced costs of operation and ownership, a smooth evolution to packet services as required, and increased trunk capacity. Additionally, the switch is largely future-proof, fully capable of handling evolving applications.

These manufacturers understand the direction that their clients are going and also understand the demands being placed on them by *their* clients. They are building products that satisfy the requirements of the third tier, their customers' customers. They are also helping service providers create the network that will position them to make the leap to being true *service providers*.

SUMMARY

In this section, we dissected the network cloud and its access extremities. We started with a discussion of access technologies, including 56K modems, ISDN, cable modems, wireless solutions, and DSL. Once we selected a method for accessing the cloud, we examined the technologies that make up the cloud itself: circuit, packet, and fast packet switching, including SONET, DWDM, frame relay, and ATM.

Customers rely on these technologies in various combinations to resolve business problems. They are chosen based on bandwidth requirements, cost, their capability to provide required levels of service quality, local availability, and the degree to which they will interoperate with other technologies and applications.

By and large, customers are not in the technology business and don't want to be. They have their own business issues to deal with and do not care about selecting the best technology mix to manage those issues. What they want is for the service provider to do it—to understand their business challenges, to offer a solution, and to not only have it work, but to know without a second thought that it is the *best* solution.

The key to providing that best solution lies in the ability to provide services. It is easy to believe that services come in a limited number of flavors like voice, video, image, and data transport. If a service provider believes that their business is to simply provide access and transport, then they will be remanded into the corner with the other unimaginative commodity providers. The fact is that customers can buy their access and transport services from a broad spectrum of suppliers today, and recent legal rulings not only make that possible; they make it *easy*. Commodity providers win the game if they have the cheapest price for their commodity among all the other commodity providers. Perhaps I'm naive, but that doesn't sound like a "service" to me. Dictionaries define "service" as "an act of assistance or benefit to others." Service providers, then, should be providing benefits, and in the eyes of customers, the best benefit they could possibly deliver is some form of competitive advantage. Please understand, access and transport are critical components of modern business today. Without the network, *there is no business*. However, imagine the power of a full service network that not only provides access and transport, but also has the ability to adapt to customer business requirements. Now *that's* a service.

In the next section, we will examine the convergence of companies, the second component of the convergence troika. As we observed in the introduction, there is something of a feeding frenzy underway as companies in the industry work to reinvent themselves to become the *first best provider* of services for their customers. They realize that they do not have everything it takes to do it all and also understand that they do not have the time nor the wherewithal to develop additional capabilities in-house. Were they to try, the market would leave them behind in a cloud of dust. Instead, they combine their efforts with those of other providers, thus rounding out their collective capabilities. In this next section, we will describe that phenomenon.

Finally, in the last section of the book, we will discuss the basic services that customers want, including voice, video, image, data, fax, and e-mail, but we will not stop there. Service convergence means logically integrating those basic capabilities to create new and exciting applications that use the underlying technology to innovatively resolve business challenges. That integration process requires other technological components including signaling systems, QoS protocols, and network management. The resulting applications are powerful and compelling, and have a significant impact on the businesses that implement them. They include IP voice, unified messaging, *virtual private networks* (VPNs), and voice-enabled Web sites.

Remember, knowledge about a customer's business processes leads to the development of solutions through technology convergence. Technology convergence in turn leads to company convergence, as competitors vie to have the most complete solution in as short a timeframe as possible. Services convergence, the result of the first two, represents the fruits of the labor of technology and company convergence. And that's the end game.

BIBLIOGRAPHY

BOOKS

AT&T Bell Laboratories Technical Staff and Technical Publication Department. *Engineering and Operations in the Bell System*. Murray Hill, NJ; 1983.

Brooks, John. *Telephone: The First Hundred Years*. Harper & Row; New York, 1975.

Christensen, Clayton. *The Innovator's Dilemma*. Harvard Business School Press; Boston, 1997.

Clarke, Arthur C. *How the World Was One*. Bantam Books; New York, 1992.

Flanagan, William A. *The Guide to T-1 Networking: How to Buy, Install and Use T-1, from Desktop to DS-3*. Telecom Library, Inc.; New York, 1990.

Goralski, Walter. *ASDL and DSL Technologies*. McGraw-Hill; New York, 1998.

Goralski, Walter. *SONET*. McGraw-Hill; New York, 1997.

——. *International Telecommunications Union Telecommunications Standardization Sector ITU-T G.721*. Geneva, 1996.

Kessler, Gary and Peter V. Southwick. *ISDN*. McGraw-Hill; New York, 1998.

Loshin, Pete. *TCP/IP For Everyone*. AP Professional; Chestnut Hill, MA, 1995.

Minoli, Daniel. *Telecommunications Technology Handbook*. Artech House; Norwood, MA, 1991.

Newton, Harry. *Newton's Telecom Dictionary*. Miller Freeman; New York, 1999.

Stallings, William. *Data and Computer Communications, 6th Edition*. Prentice-Hall; Upper Saddle River, NJ, 2000.

WEB RESOURCES

Dense Wavelength Division Multiplexing. From ATG's Communications & Networking Technology Guide Series, sponsored by Ciena. Available at www.techguide.com.

Heywood, Peter. "Doing in ATM?" Data Communications, available at www.data.com/issue/990421/doing.html.

HomePNA Specification 1.0 Field Tests Status. Available at the Home Phoneline Networking Alliance Web site, www.homePNA.com, June 1998.

Parente, Victor R. "Packet Over SONET: Ringing Up Speed." Available at www.data.com/tutorials/sonet.html.

"Simple, High-Speed Ethernet Technology for the Home." A white paper published at the Home Phoneline Networking Alliance Web site, www.homePNA.com, June 1998.

"Voice Over Frame Relay: A technical brief by the Frame Relay Forum." Available at www.frforum.com/4000/voicetechbrief.html.

ARTICLES

The ATM Forum. "Unleash the Power: Building Multi-Service IP Networks with ATM Cores." Available from the ATM Forum Web site, www.atmforum.com.

Barrett, Randy, and Carol Wilson. "Trouble in DSL's Broadband Paradise." *Interactive Week*, October 25, 1999.

Blake, Pat. "Will Carriers Bypass SONET and Take Data to DWDM?" *America's Network*. Special supplement.

Broderson, Mikkel and Bengt Beyer-Ebbesen. "ATM Flow Control: Recipe for Demanding Apps." *Network World*, August 31, 1998, page 29.

Bosch Telecommunications. "LMDS—Architectural Overview." From the corporate Web site, www.boschtelecominc.com/lmds/architec.htm.

Carlson, Randy and Martyn Roetter. "Justifying the Need for Fiber Networks." *America's Network*, February 15, 1999, page 28.

Caruso, Jeff. "Swedish Trio Touts ATM Alternative." *Network World*, January 18,1999.

Christopher, Abby. "Now, Play Nice." *Upside*, October 1999, page 100.

Clark, David S. "High-Speed Data Races Home." *Scientific American*, October 1999, page 92.

Clarke, Arthur C. "Extra-Terrestrial-Relays: Can Rocket Stations Give World-Wide Radio Coverage?" *Wireless World*, October 1945, pages 305-307.

Collins, Greg and Tam Dell'Oro. "Gigabit Ethernet—The Market Takes Off." *Business Communications Review*, April 1999, page 32.

Coursey, David. "Data, Data Everywhere." *Upside*, September 1999, page 148.

Crockett, Roger O. "Why Motorola Should Hang Up on Iridium." *Business Week*, August 30, 1999, page 46.

Duffy, Jim. "Emerging Standard to Speed Up Ethernet Reconfigs." *Network World*, November 1, 1999, page 8.

Federal Communications Commission. "FCC Action to ACCELERATE availability of ADVANCED telecommunications SERVICES for residential and small business consumers." *Common Carrier Action*, November 18, 1999.

Galitzine, Greg. "The Wireless Last Mile." *Internet Telephony*, June 1999, page 60.

Globalstar Corporation. "About Globalstar." From the corporate Web site, www.globalstar.com/en/about/index.html.

Greene, Tim. "A New Twist on DSL: Voice Services." *Network World*, August 16, 1999, page 36.

———. "HDSL2 Could Mean Cheaper T-1s for You." *Network World*, October 4, 1999, page 6.

———. "QWEST and IXC Join Big Boys with DSL Offerings." *Network World*, August 9, 1999, page 8.

———. "Splitterless DSL Promises Faster Service Delivery." *Network World*, November 1, 1999, page 12.

Heywood, Peter. "Doing in ATM?" *Data Communications*, April 21, 1999, page 67.

"IP/ATM Platform Tackles Brain as well as Brawn." *America's Network*, February 15, 1999, page 14.

Korzeniowski, Paul. "Local-Loop Products Are Good for the Long Haul." *Internet Week*, September 13, 1999, page 48.

———. "Sun Setting on FDDI." *Business Communications Review*, April 1999, page 47.

Langdon, Greg. "Voice Over DSL Sounds Promising." *Network World*, August 2, 1999, page 31.

Lindstrom, Annie. "Fiber 'Firsts' Shine Bright." *America's Network*, July 15, 1999, page 26.

———. "SONET Celebrates its 15th Anniversary." *America's Network*, September 1, 1999, page 32.

———. "The Fiber Webmasters." *America's Network*, December 1, 1998, page 19.

———. "The Opti-mologists." *America's Network*, September 1, 1999, page 20.

Lucent Technologies. "Lucent Technologies Launches Breakthrough DSL Platform to Deliver High-Quality Voice, Data, and Video Services." September 7, 1999. Company press release.

Makris, Joanna. "One-Pipe-Fits-All ATM." *Data Communications*, July 1998, page 26.

Mason, Charles. "Competition from Above?" *America's Network*, July 15, 1999, page 16.

———. "Keeping on Track with 3G." *America's Network*, September 15, 1999, page 61.

———. "The Long and the Short of 3G." *America's Network*, May 1, 1999, page 38.

Matsumoto, Craig. "Optical Nets Could Threaten IP's Future." *Data Communications*, August 20, 1999.

Moozakis, Chuck. "High-Speed Token Ring Moves Closer to Reality." *Internet Week*, March 25, 1998.

Moran, Rich and Steve Cortez. "Evolving Toward the All-Optical Network." *America's Network*, September 1, 1999, page 39.

Mosquera, Mary. "DSL High-Speed Rollout to Snowball in 2000." *TechWeb*, September 22, 1999.

"The North American Fiber Highway System." *America's Network*, February 15, 1999, page 28.

Olicom Corporation. "Olicom Sells Token Ring Business to Madge Networks." August 31, 1999. Company press release.

Panditi, Surya. "The Power of Light." *Communications News*, July 1999, page 24.

Pappalardo, Denise. "Cisco, Motorola Buy into Wireless Broadband." *Network World*, June 14, 1999, page 14.

Pearce, Alan. "An Unexpected Turn for 3G." *America's Network*, May 15, 1999, page 58.

Rohde, David. "All-in-One-Access." *Network World*, September 27, 1999, page 77.

Rysavy, Peter. "Broadband Wireless: Now Playing in Select Locations." *Data Communications*, October 1999, page 73.

Slater, Bill. "Have it Your Way!" *Communications News*, August, 1999, page 74.

Takahashi, Dean. "Start-Up Extends Reach of DSL Lines." *Wall Street Journal*, October 14, 1999, page B8.

Teledesic Corporation. "Teledesic Fast Facts." From the corporate Web site, www.teledesic.com/about/about.htm.

Tolly, Kevin. "One Giant Step for High-Speed Token Ring." *Network World*, June 1, 1998.

Turner, Brough. "The Impact of CTI on Wireless Communications." *CTI*, May 1999, page 44.

Willis, David. "The Year of the ATM WAN?" *Network Computing*, July 26, 1999.

Zimmerman, Christine. "ATM! The Technology That Would Not Die!" *Data Communications*, April 21, 1999, page 46.

COMPANY CONVERGENCE

Probably the most important management fundamental that is being ignored today is staying close to the customer to satisfy his needs and anticipate his wants. In too many companies, the customer has become a bloody nuisance whose unpredictable behavior damages carefully made strategic plans, whose activities mess up computer operations, and who stubbornly insists that purchased products should work.

> Tom Peters and Robert Waterman.
> *In Search of Excellence.*

If HP knew what HP knows, we would be three times as profitable.

> Lew Platt, former CEO of HP, from
> Thomas H. Davenport and Laurence Prusak's
> *Working Knowledge.*

In Part One, we discussed how knowledge about the customer helps network providers select the technologies that will help them satisfy those customers' requests for service. We also discussed the fact that technology convergence is only a piece of a rather complex puzzle. The second piece, company convergence, must occur in lockstep if service convergence is to take place.

When service providers assess their roles in the modern telecommunications marketplace, they often discover that the game has changed so radically that they don't know the rules anymore. Customer expectations have shifted, lines of business have become blurred, players have changed, and the roles that were so well defined for each of them seem foreign.

The tradition-bound model of manufacturer selling to wholesaler, wholesaler to retailer, and retailer to consumer is rapidly disappearing as e-commerce and e-business change the model of doing business, and with it the competitive landscape. The stately, linear business processes that have characterized corporate practices for a very long time are giving way to faster, sleeker models, driven by a demanding customer base that is not afraid to exercise its right of competitive choice. The Internet, perhaps the greatest change agent in business history, is redefining the market and forcing a redefinition of the business of doing business.

E-COMMERCE

Nothing is remarkable about the process of making an online purchase. In fact, it is not all that different from making a purchase by telephone. Other than the fact that the telephone agent is replaced with a Web server that performs the same functions, the process is identical. Yet e-commerce has become one of the most influential phenomena of the Internet era.

So what is it exactly? Opinions differ greatly on a precise definition. During a Senate committee hearing on pornography associated with the Communications Act of 1996, one of the ranking senators on the committee was asked to state his own definition of pornographic material. After careful thought and a certain amount of fluster, he responded, "Look, I can't define it, but I know it when I see it." E-commerce is equally hard to define, but when implemented, its impact is as hard as stone. To many, e-commerce consists of the process of ordering merchandise by phone from a mail order company, sending a fax to the diner next door to order lunch, or ordering a pizza from Pizza Hut (www.pizzahut.com). To others, it is the use of *electronic data interchange* (EDI) transactions between automobile manufacturers and their vendors to carry on the business of manufacturing cars, or a process by which a service provider sends an electronic bill to a large customer in lieu of a paper invoice.

All of these are versions of e-commerce, but today it is usually taken to mean the process of (1) shopping online, (2) selecting a product, (3) securely paying for the product, and (4) ensuring that the customer receives their purchase within a reasonable amount of time. In all cases, it involves the use of secure payment protocols to protect the buyer and in some cases may involve the use of electronic "money" to protect their anonymity.

WHY E-COMMERCE?

Numerous factors have led to the success of e-commerce, including convenience, lower cost, access to global product markets, and anonymity. Its success is also founded upon compelling economic reasons. A June 1999 study funded by Cisco and conducted by the University of Texas demonstrates that online commerce is not a passing fad. In 1998 alone, the Internet economy generated $301 billion in U.S. revenues and was directly responsible for 1.2 million jobs. It also indicated that Internet workers are, on average, 65 percent more productive than non-Internet workers. Furthermore, the average revenue per Internet employee was approximately $250,000 compared to $160,000 for the non-Internet employee. At $300 billion, the Internet economy rests in the same lofty heights as the automotive industry ($350 billion) and the telecommunications market ($270 billion).

In a separate study, Forrester Research observed that in 1998 business-to-business e-commerce was a $43-billion industry but will climb to an almost unimaginable $1.3 trillion by 2003.

FROM E-COMMERCE TO E-BUSINESS

E-commerce is part of a much larger phenomenon called *e-business*. E-business involves moving key business processes to the Web in order to gain efficiency, speed, and marketshare. According to consultancy organization V-One, the move to e-business will result in a 60 percent savings in network costs for many corporations as they migrate away from leased facilities in favor of public networks that have the ability to emulate the services provided by a private-line connection. The *Automotive Network Exchange* (ANX), a complex network that serves the automotive industry, is made up of dedicated facilities but is in the process of being redesigned as a *Virtual Private Network* (VPN) to eliminate the cost of leased facilities without sacrificing the safety and security that they provide.[1]

The retail industry has jumped on the e-business bandwagon with gusto as well. Consider the following scenario. A large retailer with hundreds of outlet stores and many product suppliers implements e-business processes to make their business more efficient. As part of the process, they use decremental inventory control systems to track on-hand inventory of every store in their chain in real time. Several times a day, inventory information is transmitted to headquarters, where the data is archived and used for sales report generation.

To ensure that their suppliers can anticipate demand and respond quickly to inventory requests, the retailer's internal computer systems are logically connected to those of each major supplier. When a product from a particular supplier in any store hits a "low watermark" in the inventory system, an order is automatically transmitted from the retailer to the supplier. The supplier responds by sending a confirmation and invoice to the retailer, as well as a delivery order to the shipping department with instructions to add that particular product to the outgoing shipment for the store. Meanwhile, the retailer's backroom systems automatically pay the bill using an accepted online payment protocol. Note that the only human involved in

---

[1]See the section on VPNs later in this chapter.

this process is the person who drives the forklift on the loading dock and loads the product on the truck to be delivered. Perhaps Warren Bennis, Professor Emeritus at the University of Southern California, was right when he proclaimed that "the business of the future will be run by a person and a dog. The person will be there to feed the dog; the dog will be there to make sure that the person doesn't touch anything."

Some retailers are even more innovative. One major chain combines sales information from each store with local data feeds from the National Weather Service. By passing the sales data and weather information through the digital equivalent of a food processor, they can derive accurate predictive algorithms that enable them to anticipate weather changes and therefore ensure that they have the right products on hand when the weather changes. This takes just-in-time inventory to a whole new level!

CORPORATE EVOLUTION

To understand the behavior of the modern corporation, it is useful to examine how it evolved from its industrial roots and how the defining characteristics have changed. These include

- The management model that guides the corporation
- The internal corporate structure
- The relationship between the corporation and its peers and competitors
- The nature of the value chain
- The customer's perception of the corporation and its role
- The definition of value in the corporation

Modern corporations have their roots in the industrial age, when corporations relied on the mass production of a commodity to achieve market supremacy. These corporations were largely supplier-driven, because their ability to create products was based entirely on the availability of some scarce physical resource. Furthermore, these industries tended to be vertical in nature, creating a single homogeneous product that was in widespread demand across a variety of *other* vertical industries. Consider the steel business, for example. During the industrial age, steel was king. The industry, however, was dependent upon a steady supply of ore and personnel, and as long as there was an abundance of both, it prospered. Unfortunately, operational inefficiencies and personnel costs did prosper in the American steel industry as offshore operations figured out ways to do the same job with less people and therefore at significantly lower cost. Today there is still a steel industry in the U.S., but it is a different business from the one it replaced, manufacturing small volume lots of highly specialized, custom-designed items. The focus is on customization and service, not product volume.

Businesses can be characterized by two major indicative characteristics. One is the degree to which they innovatively create value in their product to distinguish them from their competitors. The other is the nature of the resources they use to create products that they sell. If the corporation analyzes their market well and converts the raw material of their business into innovative product sets, they will succeed and prosper.

Let's consider the business characteristics we discussed earlier as they relate to the industrial age corporation. We identified six of them:

The management model that guided the corporation was a top-down, hierarchical, and rigid structure. All authority was concentrated in the upper tier of the management hierarchy; knowledge was highly compartmentalized.

The internal corporate structure was equally rigid, shaped like a pyramid with multiple levels of management that expanded in number as the levels descended.

The relationship between the corporation and its peers and competitors was somewhat belligerent. Because they sold commodities, price was the only differentiator they typically had. Therefore, they aggressively competed with each other, sometimes in less than business-like ways.

The nature of the value chain was absolutely linear. On one end of the business, raw materials were delivered in trucks. At each progressive step of the manufacturing process, an incremental bit of value was added. Finally, at the end of the process, steel was the result. This is an example of the classical "fishbone diagram" that illustrates the linear nature of legacy industries.

The customer's perception of the corporation and its role was one of awe. These corporations were all-powerful and in most cases dictated terms to the customer.

The definition of value in the corporation was based on the degree to which they could hold sway over their suppliers, and the degree to which they could out-produce the competition. The name of the game was volume.

As corporations evolved in the 1940s, '50s, and '60s, things changed, but interestingly enough many of the aforementioned business characteristics did not. Vertical industries evolved (the so-called smokestacks or stovepipes), but many of them, including the telephone companies, continued to rely on the management models of the industrial age. They used a "divide and conquer" approach, believing that the best way to ensure that a complex task would be completed was to break the task into myriad subtasks, and then to compartmentalize the knowledge required to complete each task.[2] In this model, workers were directed to become highly specialized, to focus on their particular subtask, and to not be concerned with the greater task at hand. If the process required more speed, the corporation simply threw more resources at it in the form of people, which constituted an "inexhaustible resource" in many corporations. This model continued to be applied until well into the 1980s in many corporations, including telecommunications.

Today the model is quite different, with the principal goals being to improve operational efficiencies wherever possible and

---

[2]Consider the role of the *Bell System Practices* (BSPs). These books documented every possible task that could ever be undertaken in a telephone company, broken down to an incredibly detailed degree.

to greatly reduce costs. Furthermore, the six managerial characteristics have changed dramatically. The top-down, hierarchical, centrally controlled corporation has been replaced with a flatter organizational structure with distributed decision-making power at all levels. Instead of adversarial relationships with their competitors, many corporations have crafted cooperative relationships that ensure the survival of all players. The term "coopetition" that describes this "cooperative competition" concept came out of this evolving relationship.

The value chain has changed as well. Instead of being linear and stately, it has become completely non-linear and moves rather fast. *Value chains* have become *hyperchains,* and customers now perceive the corporation as a peer that is involved as a partner in the success of their business. Furthermore, a value chain of information has been defined that does not deal with the processes involved in the manufacture of a finished product, but with the development of knowledge necessary to ensure that it is the *right* finished product.

Perhaps the single most important change is the manner in which corporations perceive value. Industrial age corporations measured value according to their ability to produce massive quantities of products. Internet age corporations measure value according to how well they use the knowledge they acquire about their customers, suppliers, and competitors to create advantageous positions for themselves and use that position to provide superior customer service. In the Internet age, individuals work in specialized teams rather than on their own. Those teams may include members from various organizations within the company, from other companies, and even customers. Today, for example, it is not uncommon for an *Incumbent Local Exchange Carrier* (ILEC) sales team and an equipment manufacturer sales team to jointly call on a customer. As a result of this collaborative teaming approach, the focus is on the greater task, rather than on the subtask. The team becomes much more future-focused, significantly more responsive to customer requests, and therefore more competitively positioned. In these corporations, competition exists *not* between individual corporations but rather between *clusters of companies.* It becomes a matter of survival for businesses to form tight relationships with consumers, suppliers, retailers, distributors, and competitors.

THE IMPORTANCE OF CORPORATE KNOWLEDGE

In the new marketplace, knowledge-based corporations rise to the top of the competitive heap because they have learned that knowledge about the customer is the single most important resource they have. It isn't enough to have the best router, the most bandwidth, or the most survivable network. Unless the "ore" is manipulated in some way to make it more targeted at each customer's business challenges, it's just more ore.

In a prior life (before telecommunications), I was a scuba diving instructor. One of the skills that all students are required to learn is buddy breathing, the process by which two divers share one regulator. Buddy breathing is used whenever (1) a diver has an out-of-air emergency and (2) there is no clear path to the surface, as might happen if the diver were to run out of air while inside a shipwreck, in a cave, or under a kelp canopy. It turns out that buddy breathing actually has a lot to do with the proper use of corporate knowledge, so bear with me. Besides, it's an entertaining break from technology.

Buddy breathing is an inherently difficult exercise. If a diver finds that they are out of air, they first have to find their buddy, not typically an easy thing to do. Next, they must communicate to their buddy that they are out of air, a process that is accomplished through hand signals that are not intuitive ("What? You found a treasure ship?"). Once the buddy understands the plight the other diver is in, the two must position themselves to begin buddy breathing. Because the second stage of the regulator (the part that goes in your mouth) always comes over the right shoulder, the person requesting the air must be on the donor's left side. Once they have managed to get into the appropriate positions, they can begin to share the donor's air by passing the second stage back and forth.

This, of course, is where the whole thing usually comes unraveled. By the time the diver realizes that he is out of air, finds his buddy, communicates that he needs to buddy breathe, gets into the right position, and finally starts sharing air, several minutes have passed, after which time the whole process becomes somewhat academic. Even if the process happens without a hitch, the donor naturally assumes that the person requesting the air *is going to give the regulator back*. Because of these minor problems, I always told my students that in the unlikely event that they found themselves unable to breathe because of an equipment problem or because they ran out of air, they should not swim up to me and try to communicate with me using hand signals. The only hand signals I wanted to see, I told them, were those used to pull the regulator out of my mouth and place it in theirs. I'll understand, I told them. I have another one.

So what does this have to do with corporate knowledge? A lot, as analogies go. A diver should never under *any* circumstances run out of air, because on the diver's left side is a pressure gauge that indicates in real time exactly how much air remains in the tank. The pressure gauge, however, only works if you look at it occasionally.

Every night, corporations carefully make copies of their transaction databases. They store the tapes in sealed cases that are picked up the next day by an archival storage company that transports the tapes to a sealed vault where they will reside forever. They are retrieved if a database recovery be required due to a disk failure, but otherwise the information on those tapes stays archived.

Those tapes are like the diver's pressure gauge. They contain invaluable information about customer buying patterns, product return information, and other data that can be converted to competitive advantage. However, unless the corporation distills that information out of their databases, it is useless to them.

ENTERPRISE RESOURCE PLANNING (ERP)

The need to collect, analyze, and respond to knowledge about customers is of paramount importance today, because, if executed properly, knowledge management can become the most critical competitive advantage a company has. This process of collecting the data, storing it so that it can be analyzed, and making decisions based on the knowledge it provides falls under a general family of processes called *Enterprise Resource Planning* (ERP).

ERP is best defined as a corporate planning and internal communications system that widely affects corporate resources. ERP systems are designed to address such functions as planning, transaction processing, accounting, finance, logistics, inventory management, sales order fulfillment, human resources operations, payroll, and customer relationship management. It serves as the umbrella function for a variety of closely integrated functions that are characteristic of the knowledge-based corporation.

ERP's functional ancestor was an application called *Materials Requirement Planning* (MRP), developed in the 1970s to assist manufacturing companies with the difficult task of managing their production processes and natural resources procurement. MRP systems generated production schedules based on currently available raw materials and alerted management when raw materials needed to be restocked.

Over time, MRP evolved into *Manufacturing Resource Planning* or *MRP II*. In addition to the tracking and alerting functions of MRP, MRP II enabled production managers to create "what if" scenarios that helped them plan for unprecedented events such as a shortage of raw materials or a late start on another project, the delayed output of which would affect the project at hand.

ERP takes the evolution to the next level by integrating financial and human resources concerns into the solution. Unlike MRP and MRP II, which originated in the production industry, ERP focuses more on the business side of the enterprise, concerning itself with the internal operational activities of the company.

THE ERP PROCESS

The process by which ERP yields business intelligence folds nicely into the overall convergence hierarchy. During a typical business day, interactions with customers yield enormous quantities of data that are stored in corporate databases. The data might include records of sales, product returns, service order processing, repair data, notes from meetings with customers, competitive intelligence, supplier information, and so on, all the result of normal business activities.

ERP defines an umbrella under which are found a number of functions including *data warehousing, data mining, knowledge management,* and *customer relationship management* (CRM), as shown in Figure 2-1. ERP outlines the process by which corporate data is developed into a finely honed competitive advantage.

FIGURE 2-1   The ERP umbrella

This corporate data is typically archived in a database (or multiple databases) within a data warehouse of some sort, typically a disk farm behind a large processor in a data center. At this point, the data is an unstructured collection of business records stored in a format easily digestible by a computer that does not yet have a great deal of strategic value. In order to gain value, the data must somehow be manipulated into *information*. Information is defined as a collection of facts or data points that have value to a user.

The raw material of information is typically accumulated using a technique called *data warehousing*. Data warehouses store data but have little to do with the process of converting it to information. Data warehouses make the data more available to the applications that will manipulate it but do not take part in the process of converting the data into business intelligence. In fact, according to Brio Corporation, corporate information systems and the users who interact with them form an "information supply chain" that includes the raw material (data), the distribution systems themselves (the data warehouse and corporate network), the manufacturers or producers (the information technology organization), and the consumers (end users).

DATA MINING.   To convert data into information, a process has been created called *data mining*. Data mining is the technique of identifying, examining, and modeling corporate data to identify behavior patterns that can be used to gain a competitive business advantage. Gartner Group defines data mining as "the process of discovering meaningful new correlations, patterns, and trends by sifting through large amounts of data stored in repositories using pattern recognition technologies as well as statistical and mathematical techniques." Aaron Zornes of the META Group defines it as "a knowledge discovery process of extracting previously unknown, actionable information from very large databases."

Data mining came about because of the recognized need to manipulate corporate databases in order to extract meaningful information that supports corporate decision-makers. The process produces information that can be converted into *knowledge*, which is nothing more than a sense of familiarity or understanding that comes from experience or study. Knowledge yields competitive advantages.

Data mining relies on sophisticated analytical techniques such as artificially intelligent filters, neural networks, decision trees, and analysis tools that can be used to build a model of the business environment based on data collected from multiple sources. It yields patterns that can be used to identify and go after new or not-yet-emerged business opportunities. Corporations that implement a data mining application do so for many of the same reasons, including the following:

- *Customer and market activity analysis*   By analyzing the buying and selling patterns of a large population of customers, corporations can identify such factors as what they are buying (or returning), why they are buying (or returning), whether or not their purchases are linked to other purchases or market activities, and a host of other factors. One major retailer claims that their data mining techniques are so indicative of customer behavior that, given certain conditions in the marketplace that they have the ability to predict, they can dramatically change the sales of tennis racquets by lowering the price of tennis balls by as little as a penny.

- *Customer retention and acquisition*   By tracking what customers are doing and identifying the reasons behind their activities, companies can identify the factors that cause customers to stay or go. Domino's, for example, carefully tracks customer purchases, relying on Caller ID information to build a database of pizza preferences, purchase frequencies, and the like. Customers enjoy the personalized service they get from the company when they place orders. Similarly, Domino's can identify customers who ordered once and never ordered again, and send them promotional material in an attempt to regain them as a customer.

- *Product cross-selling and upgrading*   Many products are complementary, and customers often appreciate being told about upgrades to the product they currently have or are considering buying. Amazon.com is well known for providing this service to their customers. When a patron has made a book purchase online, they often receive an e-mail several days later from Amazon, telling them that the company has detected an interesting trend; people who bought the book that the customer recently purchased also bought other titles in greater than coincidental numbers. They then make it inordinately easy for the customer to "click here" to purchase any of the other books. This is a good example of the benefit of data mining.

· *Theft and fraud detection*   Telecommunications providers and credit card companies may use data mining as a way to detect fraud and assess the effectiveness of advertising. For example, American Express and AT&T will routinely call a customer if they detect unusual or extraordinary usage of a credit or calling card, based on the customer's known calling patterns. The same information is often used to match a customer to a custom-calling plan.

A common theme runs through all of these subtasks. Data mining identifies customer activity patterns that help companies understand the forces that motivate customers. The same data helps the company understand the vagaries of their customers' behavior to anticipate resource demand, increase new customer acquisitions, and reduce the customer attrition rate. These techniques are becoming more and more mainstream. Gartner Group predicts that the use of data mining for targeted marketing activities will increase from less than five percent to more than 80 percent over the course of the next 10 years. And from a financial perspective, the META Group predicts that revenues from data mining will grow to as much as $800 million by the end of 2000. A number of factors have emerged that have made the "mainstreaming of data mining" possible, including improved access to corporate data through extensive networking and more powerful, capable processors, as well as statistical analysis tools that are accessible by non-statisticians because of *graphical user interfaces* (GUI) and other enhanced features.

A significant number of companies have arisen in the data mining space, creating software applications that facilitate the collection and analysis of corporate data. One challenge that has arisen, however, is the fact that many data mining applications suffer from integration difficulties, don't scale well in large (or growing) systems, or rely on a limited number of statistical analysis techniques. Consequently, users spend far too much time manipulating the data and far too little time analyzing it.

Another factor that often complicates data mining activities is the corporate perception of what data mining does. Far too often, decision-makers come to believe that data mining alone will yield the knowledge required to make informed corporate decisions, a perception that is incorrect. Data mining does not make decisions; people do, based on the *results* of data mining. It is the knowledge and experience of the analysts using data mining techniques that convert its output into usable information. So, Professor Bennis is only partially correct.

To work properly, then, data mining software must rely on multiple analytical techniques that can be combined and automated to accelerate the decision-making process and that deliver the results in a format that makes the most sense to the user of the application. A number of companies have entered the data mining game including SAS Institute, which offers a comprehensive data mining solution that includes both software and services, and Hewlett-Packard, which has enjoyed a collaborative relationship with SAS since 1997.

KNOWLEDGE MANAGEMENT. Data mining yields information that is manipulated through various managerial processes to create knowledge, the golden elixir of customer service. Data mining is an important process, but the winners in the customer service game are those capable of collecting, storing, and manipulating knowledge to create indicators for action. Ideally, those indicators are shared within the company among all personnel who have the ability to use them to bring about change. According to the *Mercer Marketplace 2000 Survey*, 80 percent of those responding believe that the process of transforming information into knowledge is the second most important source for competitive advantages, after customer relationship management.

The collection of business intelligence is a significant component of knowledge management and yields a variety of benefits, including improvements in personnel and operational efficiencies, enhanced decision-making capabilities, more effective responses to market movement, and the delivery of innovative products and services. The process also adds certain less tangible advantages, including greater familiarity with the customer's business processes as well as a better understanding of the service provider on the part of the customer. By combining corporate goals with a solid technology base, business processes can be made significantly more efficient. Finally, the global knowledge that results from knowledge management yields a more "Zen-like" understanding of business operations.

OBSTACLES TO EFFECTIVE KNOWLEDGE MANAGEMENT. Knowledge, while difficult to quantify and even more difficult to manage, is a strategic corporate asset. However, because it largely exists "in the heads" of the people who create it, an infrastructure must be designed where knowledge can be stored, maintained, and archived in a way that makes it possible to deliver it to the right people at the right time, always in the most effective format for each user. The challenge is that knowledge exists in enormous volumes and grows according to Metcalfe's Law. Bob Metcalfe, co-founder of 3Com Corporation and co-inventor of Ethernet, postulated that the value of information contained in a network of servers increases as a function of the square of the number of servers attached to the network. In other words, if a network grows from one server to eight, the value of the information contained within those networked servers increases 64 times over, not eight. Clearly, the knowledge contained in the heads of a corporation's employees multiplies in the same fashion, increasing its value exponentially.

However, this observation has practical limitations. According to Daniel Tkach of IBM, the maximum size of a company where people actually know each other and who also have a realistic and dependable understanding of the collective corporate knowledge is somewhere between 200 and 300 people. As the globalization of modern corporations and the markets they serve continues, how can a corporation reasonably expect to remain aware of the knowledge they possess, when that knowledge is scattered to the four corners of the planet? Studies have indicated that managers glean two-thirds of the information they require from meetings with other personnel. Only one-third comes from documents and other "non-human" sources. Clearly, there is a need for a knowledge management technique that employees can use to store what they know so that the knowledge can be made available to everyone.

According to Laura Empson, a recognized authority on knowledge management, 78 percent of major U.S. corporations have indicated their intent to implement a knowledge management infrastructure. However, because the definition of knowledge management is still somewhat unclear, companies should spend a considerable amount of time thinking about what it means to manage knowledge and what it will take to do so. Empson observes that knowledge includes experience, judgment, intuition, and values, none of which are easily codifiable.

What must corporations do, then, to ensure that they put into place an adequate knowledge management infrastructure? In theory, the process is simple, but the execution is somewhat more complex. First, they must clearly define the guidelines that will drive the implementation of the knowledge management infrastructure within the enterprise. Second, they must have top-down support for the effort with adequate explanations of the reasons behind the addition of the capability. Third, and perhaps most important, they must create within the enterprise a culture that places value on knowledge and recognizes that networked knowledge is far more valuable than standalone knowledge within one person's head. Too many corporations have crafted philosophies based upon the belief that knowledge is power, which causes individuals to hoard what they know in the mistaken belief that to do so creates a position of power for themselves. According to a survey conducted in 1997 by the

Journal of Knowledge Management, the single greatest barrier to the successful implementation of a knowledge management infrastructure is the failure of the organization to recognize the value of the new process and accept it. Fourth, the right technological substrate must be selected to ensure that the system not only works, but also works efficiently. Corporations often underestimate the degree to which knowledge management systems can consume computer and network resources. The result is an overtaxed network that delivers marginal service at best.

As corporations get larger, it ironically becomes more and more difficult to manage the processes of managing their day-to-day operations. These processes include buying, invoicing, inventory control and management, and any number of other functions. Consulting firm AMR estimates that *maverick buying*, that is, the process of buying resources in a piecemeal fashion by individuals instead of through a central purchasing agency, accounts for as much as a third of their total operating resource expenditures. This adds a 15- to 27-percent overlay premium on those purchases due to the loss of critical mass from one-off purchasing. According to studies conducted by Ariba, one of the most successful providers of knowledge management software, corporations often spend as much as a third of their revenues on operating resources such as office supplies, *information technologies* (IT) equipment and services, computers, and maintenance. Each of these resources has associated with it a cost of acquisition that can amount to as much as $150 per transaction when done using traditional non-mechanized processes. Systems like those from companies such as Ariba can help corporations manage their costs by eliminating the manual components in favor of computer and network-driven processes.

Ariba's *Operating Resource Management* (ORM) application brings order to the knowledge management chaos through the implementation of an electronic infrastructure that enables corporations to strategically monitor and manage their operating resources. By taking advantage of technology, ORM lowers transaction costs through the use of e-commerce and decision support system automation. By concentrating on the ORM process, corporations can build strategic relationships with suppliers and use the power of aggregate buying to achieve volume discounts.

Another major player in the knowledge management game is SAS. The company's Collaborative Business Intelligence product includes enterprise reporting, data warehousing, customer relationship management, and data mining. This combination of processes unites numerical business intelligence information with text-based, unstructured content. By bringing the two together, businesses can share, distill, and reuse knowledge that would previously have been difficult to locate.

SAS' Collaborative Business Intelligence application offers users the ability to search for information in a variety of ways as well as an innovative subscription mechanism that enables them to receive unsolicited notification when information changes.

Another player in the business intelligence game is Lotus, currently offering a package that is integrated with the company's Domino messaging and collaboration products. The software includes a knowledge management portal that enables customers to organize specific information according to their individual interests and organizational responsibilities. Users can personalize the portal to provide e-mail, calendar services, discussion areas, and quick access to preferred Web sites. The application also includes a powerful and customizable search engine that enables employees to search according to personal search profiles.

Luna, another powerful player, offers applications that manage hierarchical business relationships such as discount structures, service volume requirements, and differences based on each customer's geographic location. Within the defined hierarchy, the system can further identify differences based on department and individual interactions, thus offering both gross-level as well as granular information. At the group level, Luna's products enable corporations to identify negotiated terms and conditions, deliver customized applications for each client and partner, and support collaborative business activities. At the individual level, the company's products provide all of the previous information as well as support for distinctly identifiable business roles, preferences, access levels, and personalized views of each of the identified relationships.

Several corporations offer products that are specifically targeted at the relationships between telecommunications providers and customers. Visual Networks, for example, offers a popular software and hardware combination that enables customer network managers to monitor a service provider's frame relay circuits to determine whether the provider is meeting *service-level agreements* (SLAs). Given the growing interest in SLAs in today's telecommunications marketplace, this application and others like it will become extremely important as SLAs become significant competitive advantages for service providers.

SUPPLY CHAIN ISSUES.   Many of the concerns addressed by the overall ERP process are designed to deal with supply chain issues. Although supply chains are often described as the relationship between a raw materials provider, a manufacturer, a wholesaler, a retailer, and a customer, it can actually be viewed more generally than that. The supply chain process begins the moment a customer places an order, often online. That simple event may kick off a manufacturing order, reserve the necessary raw materials and manufacturing capacity, and create expected delivery reports.

A number of companies focus their efforts on the overall supply chain, including I2. I2 offers a remarkable array of capabilities including customer service and support, relationship-building modules, system personalization, brand-building techniques, account management, financial forecasting, product portfolio planning, development scheduling, and product transition planning.

ERP IN THE TELECOMM SPACE,   ERP's promise of greatly improved operational efficiencies didn't initially attract much attention from telecom carriers, particularly the ILECs. As regulated carriers with guaranteed rates of return and huge percentages of marketshare, they did not rise quickly to the ERP challenge. Recently, however, their collective interest level has grown as the capabilities embedded in the ERP suite have become better known.

One factor that has accelerated telecom's interest in ERP is the ongoing stream of mergers and acquisitions within the telecom arena. When mergers occur, a great deal of attention is focused on their strategic implications, but once the merger has taken place and the excitement has settled down, there is a great deal of work left to do internally. Not only does the merged company now have multiple networks that must be combined, they also have multiple support systems that must be integrated to ensure seamless billing, operations, administration, provisioning, repair, maintenance, and all the other functions necessary to properly operate a large network. Telcos are inordinately dependent upon their *Operations Support Systems* (OSSs) and are therefore more than a little sensitive to the issues they face with regard to backroom system integration. One of the reasons they have been somewhat slow to embrace ERP is the fact that in many cases they have waited for ERP vendors to establish relationships with the telcos' OSS providers to ensure that together they address interoperability concerns before proceeding.

Another concern that has slowed ERP implementation among telecom providers is fear of the unknown. ERP is a nontrivial undertaking and given the tremendous focus that exists today on customer service, service providers are loathe to undertake any activity that could negatively affect their ability to provide the best service possible. For example, ERP applications tend to require significant computing and network resources if they are to run properly. Telcos are therefore reluctant to deploy them for fear that they could have a deleterious impact on service. Furthermore, ERP implementation is far from being a cookie-cutter exercise. The process is different every time, and surprises are common. The applications behave differently in different system and application environments and, as a consequence, are somewhat unpredictable. Add to that the fact that even with the best offline testing, a certain amount of mystery always remains with regard to how the system will behave when placed under an online load.

Because of the competitive nature of the marketplace in which telecom providers operate and because of the fact that the customer, not the technology, runs the market, service providers are faced with the realization that the customer holds the cards. Although the customer used to call the service provider to order a single technology-based product, today they are far more likely to come to the service provider with the expectation that they be exactly that—the provider of a full range of services, customized for the customer's particular needs and delivered in the most convenient manner.

Because of their access and transport legacy, incumbent service providers understand technology and what it takes to deploy it. As the market has become more competitive, however, they have had to change the way they view their role in the market. They can no longer rely on technology alone to satisfy the limited needs of their customers because those same customers have realized that technology is not an end itself. It is an important part of a service package. No longer can the service provider roll out a new technology and expect the market to scurry away and find something to do with it. Their focus must change from being technology-centric to service-centric. This is not bad news. The technology that they are so good at providing simply becomes a component of their all-inclusive services bundle.

CUSTOMER RELATIONSHIP MANAGEMENT (CRM). The final component in the ERP family is *Customer Relationship Management* (CRM). It is the culmination of the ERP process, the distilled vintage of information that results from an analysis of business intelligence, which in turn derives from the data mining effort.

Many companies offer CRM software today, including PeopleSoft, IBM, Siebel, Hewlett-Packard, Oracle, and others. They too are going through a merger and acquisition period as they jockey to combine their collective capabilities in a market that has suddenly awakened and is loudly demanding their services. PeopleSoft recently acquired Vantive, which makes a front-office software package that provides extremely capable service and support functions. IBM and Siebel, both major CRM providers, have formed a cross-marketing CRM relationship in which Siebel sells IBM's DB2 database and IBM sells Siebel's Relationship product. Similarly, HP and Oracle formed an alliance to jointly develop CRM software in September 1999.

Full-blown acquisitions have been in evidence as well. In October 1999, Nortel bought Clarify, Inc., which makes CRM software, while E.piphany bought RightPoint in November. E.piphany's software enables companies to collect, analyze, and respond to customer data, while RightPoint's applications add clickstream tracking, analytics, collaborative filtering, and real-time customer profiling capabilities.

Telecommunications manufacturers have also made acquisitions to add to their own capabilities. In 1999, Cisco formed a relationship with integrator KPMG, while Lucent Technologies acquired INS and, as we noted earlier, Nortel Networks acquired Clarify, Inc., a leading provider of front office, CRM, and e-business software.

PUTTING IT ALL TOGETHER. If we consider the entire ERP process, a logical flow emerges that begins with normal business interactions with the customer (see Figure 2-2). As the customer makes purchases, queries the company for information, buys additional add-on features, and requests repair or maintenance services, database entries accumulate that result in vast stores of uncorrelated data. The data is housed in data warehouses, often accessible through *Storage Area Networks* (SANs) interconnected via high-speed trunking architectures such as HyperChannel. SANs will be discussed in detail later in this chapter.

From the data warehouses, corporate data mining applications retrieve the data and massage it logically to identify trends and activities that help the company understand the drives behind customer behavior so that it can be anticipated and acted upon before the fact. The data mining process then yields information that now has enhanced value.

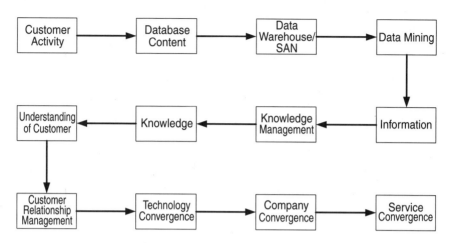

FIGURE 2-2 The logical flow of information within a knowledge-driven corporation

The information derived from data mining is manipulated in a variety of ways and combined with other information to create knowledge. Knowledge is created whenever information is mixed with human experience to further enhance its value. The knowledge can then be managed to yield an even more accurate understanding of the customer.

Once customer behavior is understood, strategies can be developed that will help the service provider anticipate what each customer will require in the future. This leads to an enhanced relationship between the company and its customers, because the company is no longer in the position of simply responding to customer requests, but in fact can *predict* what the customer will require. This represents CRM at its finest.

Consider this model within the context of the telecommunications service provider space. Once the service provider understands and has developed a relationship with the customer, and knows what the customer base is looking for in the way of products and services, the service provider can develop a technology plan that will help ensure that they have the right technologies in the right place at the right time to satisfy those service requests when they arrive. Given the accelerated pace of the telecommunications market, however, service providers do not have the time to develop new technologies in-house. Instead, they do the next best thing, which is to go to the market, identify another company that offers the technology they require, and either buy the company or form an exclusive, strategic alliance with them. The service provider gains the technological capability they require to satisfy their customer, the technology provider gains an instantaneous and substantial piece of marketshare, and the customer's requirements are not only met; they are anticipated.

This model of providing a full complement of services by a group of companies instead of a single provider is a relatively new phenomenon in business. It is often referred to as a *virtual community* and is one of the most powerful change agents in telecommunications today.

THE BIRTH OF THE VIRTUAL COMMUNITY

As corporations evolve from standalone competitive entities driven by both suppliers and the need to mass produce in order to stay competitive to knowledge-driven clusters driven by the customer and the need to create customized solutions to remain competitive, a number of qualities emerge that characterize all companies. The two most compelling of these characteristics are (1) the degree to which the corporation allows its internal entities to self-manage and (2) the amount of value the corporation adds to the product through its own actions. We will examine each of these.

In terms of managerial control, corporations range from being completely hierarchical with all decision-making power concentrated in the upper managerial echelons to totally self-managed entities where decisions are made by lower level managers on a peer-to-peer basis. Hierarchical management certainly works; it maintains tight control over all functions within the corporation such that senior managers have the ability to control the affairs of the business. A hierarchically managed corporation is effective when competitors and customers are also hierarchically managed, but tends to be seen as unresponsive and pedantic by customers or competitors that have adopted the peer-to-peer model. It is difficult to respond quickly to requests for service or to a competitive threat when all decisions must pass through a hierarchical managerial bottleneck.

Self-managing entities tend to move quicker, tend to make decisions more effectively, and tend to respond to competitive threats and customer requests more nimbly than their hierarchical associates. Their challenge is one of intra-corporate communication. Because they lack centralized authority, communications can suffer between the peer entities. Successful self-managed corporations have realized that even though they are structurally flat, ideologically they are hierarchical. Although the corporation may comprise a large number of loosely connected business "molecules" that work independently, they all work toward a common goal that is well known by all. Thus, their collective efforts serve the overall goals of the corporation (see Figure 2-3).

ADDITION OF VALUE

Value is added to a corporation's products in a variety of ways, ranging from nothing more than a place to buy them (minimal added value) to integration services that help customers position products and services to their own competitive ends. For example, an ILEC might provide nothing more than access and transport, in which case its added value is minimal. On the other hand, it might provide consulting services that help its customers select the appropriate technology solutions based on what they know about their own customers. In this case, the added value is substantial, and the corporation's value in the eyes of its customers is enhanced.

As companies evolve in the information-driven economy, the degree to which they exhibit these two characteristics allows them to be categorized according to a limited set of behavior models. These models are useful when analyzing companies,

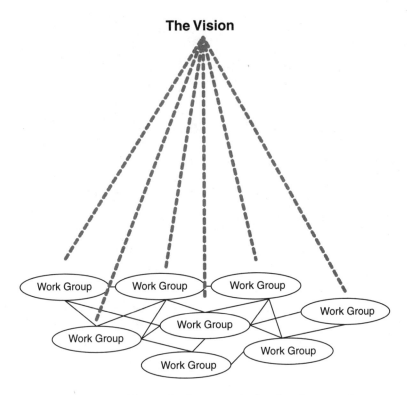

FIGURE 2-3   The virtual hierarchy that exists in the virtual corporation

because the unique combination of managerial independence and embedded value indicate the nature of the company, the services it provides, and the customers it can hope to satisfy. It also serves as a powerful self-analysis tool. If a corporation determines after analyzing itself according to this model that it falls into a less than optimal category, it can use the model to determine what it needs to do to move itself in the proper direction.

COMPANY CLASSIFICATIONS

As Figure 2-4 shows, corporations fall generally into one of four models: the flea market, the arbitrator, the process improver, or the virtual corporation. All have their place and in reality most corporations exhibit characteristics of all four, as indicated in the diagram, although one is typically dominant.

THE FLEA MARKET MODEL

Most of us have been to a flea market at one time or another whether while visiting Madrid's Rastro, Buenos Aires' Plaza San Telmo, or one of the hundreds that spring up every Sunday in the U.S. in drive-in movie parking lots. Although these markets appear to be a form of organized chaos, there is a thread of

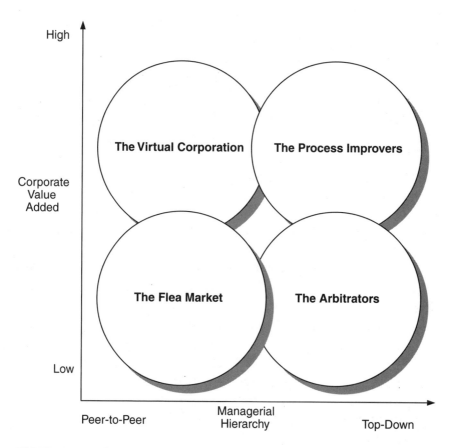

FIGURE 2-4 The corporate models

management running through them. First, the flea market comprises myriad independent, self-managed vendors, selling their merchandise side by side in a large, open market structure. Although a management overlay is there, the only added value they bring is the space they provide and the folding tables they rent to the vendors. Otherwise, they are invisible to vendor and customer alike.

In the flea market company model, several characteristics are evident. First of all, the role of buyer and seller is purely serendipitous in the sense that buyers today may well show up next week as sellers themselves. Very little control exists in this model. It is completely self-managed at an organic level. Furthermore, very little organizational trust takes place here as well. Because of the independence of each of the vendors that make up the overall market, there is minimal coordination, no hierarchical oversight, and therefore minimal trust. Caveat Emptor is the watchword.

How does market leadership fit in this model? Again, serendipity plays an active role. Market leadership has more to do with being in the right place at the right time with the right product at the right price. Leadership is fleeting in the flea market. Commodities sell well here because they are differentiated on price alone. Unfortunately, from a customer's perspective, the flea market model is not ideal. Although they may find a good price for the item they want to buy, there is no effort available to help them identify whether there might be an even better price somewhere else in the market.

Although it may not be completely obvious, flea market models are at work in (or near) the telecommunications marketplace. Online auction house eBay is a good example. eBay itself provides nothing more than a gathering place for thousands of sellers to display their wares and buyers to bid on them, hoping to secure the lowest possible price. eBay offers little added value, other than the digital equivalent of a parking lot and a folding table. Their success, of course, indicates that the service they *do* provide is a highly valued one. It is not, however, a service-differentiable organization.

An example of the flea market concept is the *Open Access Same-Time Information System* (OASIS), the online kilowatt market where power companies can auction excess capacity. Another example is Freemarkets.com, the business-to-business online auction house for buyers of industrial parts, raw materials, commodity products, and a variety of services.

Flea market businesses offer little in the way of enhanced services beyond their basic products (by design) and exhibit little coordinated management of business processes. This is by no means a criticism of businesses that model themselves this way, because many of them are quite successful.

THE ARBITRATOR MODEL

As we move up the organizational food chain, we come to the arbitrator model. Added value is still relatively low, but unlike the flea market model, here we find hierarchical management practices in use. In the arbitrator model, we find businesses that have found a market niche by placing themselves between buyers and sellers and serving as an intermediary between them. The result is a recognized market leader, because their service, while not highly differentiated, is valuable and necessary.

In many cases, arbitrators are large organizations and command significant power in the marketplaces in which they operate. As a result, they can often aggregate supply and demand by negotiating favorable prices with suppliers and then provide comfortable gathering places for customers that want to buy. In the retail world, companies like Home Depot, Costco, Staples, and Wal-Mart fulfill this role. Because of their buying power, they can create ideal market conditions for buyers and sellers, aggregating supply and demand in the most efficient manner possible.

Although the service they provide above and beyond their basic selling role is valuable, it is targeted at the mass market and is therefore not particularly specialized. Arbitrators enjoy greater trust than their flea market counterparts, because they often develop and rely on widely known brand loyalty that carries with it a high degree of trust on the part of the buyer.

Arbitrators are the best-known players in the telecommunications arena, including such names as *America Online* (AOL), E*Trade, Prodigy, and other nationally known online service providers. AOL, for example, provides a safe gathering place for buyers of a remarkable variety of products and services, while at the same time providing a gathering place for the sellers themselves. Furthermore, because of their staggering market power, they can dictate rules of behavior to create the marketplace image they desire. A recent study observed that a remarkably large percentage of AOL subscribers never leave AOL; that is, they never avail themselves of the Internet portal that is a single click away, an indicator of the degree to which AOL alone satisfies their requirements for an online service provider. As one of the company's supporters recently observed, "Think of AOL as the Love Boat. You pay one price for which you get everything. It's a totally safe place where you'll never get into any kind of trouble or run across anyone unsavory. You can go anywhere you want, and they have something for literally everyone. The problem is that you can get off the boat when it docks. And if you do, there's no telling what might happen. You could get robbed, or worse. So the best thing to do is stay on the boat!"

Arbitrators sit between buyers and sellers, bringing them together and using their considerable market power to ensure that they can offer a wide variety of products and services for the most competitive price available. They also rely on brand loyalty to ensure that they form long-term relationships with customers.

PROCESS IMPROVERS

One level up on the corporate food chain are the process improvers. Here we find a leader of the pack, who differentiates itself by improving the overall process by which its business niche operates. Such a leader tends to be higher on the added value scale, but also tends to be managed in a more top-down fashion. Instead of serving simply as a gathering place or as an aggregator of buyers and sellers, process improvers optimize the value chain to create greater value in the service they provide above and beyond the simple products or services they sell. As a result, the value they add is substantially higher than the two models we have seen previously. Automobile dealers who interview their clients and then tune preset radio stations according to their preferences are exhibiting process improvement characteristics.

Online service providers such as Amazon.com, Dell, and Cisco are examples of process improvers in the telecommunications marketplace. Amazon.com not only sells books and other products, but it also gathers customer preferences and then upsells its customers according to those preferences by sending them discount coupons and notifications of other books the company thinks the customer might like based on logged preferences. Dell interviews customers online as part of the pre-sell activity, thus ensuring that they understand exactly what the customer's requirements are before attempting to sell them a machine. Consequently, they are far more likely to satisfy the customer with their first-time purchase and keep them as a long-term customer.

Process improvers optimize the value chain, thus adding value to the products and services they sell by ensuring that doing business with them will result in the most cost-effective and efficient experience for the customer. They lead by being the best at what they do, not necessarily by having the lowest price. They count on the fact that their customer will perceive that there is inherent value in doing business with the company.

THE VIRTUAL CORPORATION

The virtual corporation is perhaps the least understood and least visible of all the business models described. Here we find that there may actually be multiple leaders, all providing specialized services to narrowly defined segments of the marketplace. The goal of the virtual corporation is to strive to understand, and perhaps even predict, specific customer challenges and provide optimized, targeted solutions for them. The virtual corporation model exhibits a phenomenon often called the convergence of money and minds, characterized by the fact that the market decides where the money should be directed within the industry, based on the fact that a particular product or service satisfies a critical demand and therefore deserves funding. Once the market's tractor beam has magnetically directed the money as it sees fit, the talented people will naturally follow the money. Consider the remarkable success of 3Com's Palm Pilot product. In quick order, the market realized how capable the device was and directed money in the direction of its manufacturer by buying Palm Pilots in the tens of thousands. Driven by the success of 3Com, talented people flocked to the corporation in droves to jump on the success bandwagon.

Corporations that exhibit the characteristics of a virtual corporation add a great deal of value to the basic products and services that they sell. They tend to be peer-to-peer management entities, relying on the abilities of individual work groups to perform necessary tasks according to some corporate vision. These work groups may form unpredictably and typically do not exist for long periods of time. They remain together long enough to accomplish a task, but then dissolve and go on to other mandates. One interesting point about these virtual entities is that they tend to cross organizational boundaries, pulling in members from the ranks of other corporate organizations, other companies, or even customers to ensure that they have the appropriate abilities at hand to accomplish the task.

A number of good examples of the virtual corporation model exist, including the *Joint Electronic Payment Initiative* (JEPI) and the *Secure Electronic Transaction Set* (SET), both made up of companies dedicated to the creation of a widely accepted set of online payment protocols. Other examples are the Java Alliance, comprising Sun, IBM, Oracle, and Netscape; Microsoft and Intel's Wintel Alliance that is designed to bolster support for the PC domain; and the Universal ADSL Working Group created by Microsoft, Compaq, and Intel, designed to bolster support for ADSL. Each of these collectives has a particular goal that they want to achieve and in some cases the members of each compete with each other.

These are, of course, "pre-packaged" in many ways, having come about because of specific long-term challenges that needed to be addressed. However, when examining the incumbent players in telecommunications today, it is an interesting exercise to determine where they each fall and ask where they *should* fall if they had the ability to change. If we examine the makeup of the market, we find seven major groups: the ILECs, the *Competitive Local Exchange Carriers* (CLECs), the *Interexchange Carriers* (IECs), the *Internet Service Providers* (ISPs) and portals, the cable providers, the equipment manufacturers, and everybody else. Each has a well-planned strategy for maintaining and growing marketshare, but the manner in which they do so varies from company to company. In the section that follows, we will examine these industry components as generic groups but will then look at specific companies to understand how they view their own roles as the marketplace evolves.

THE INCUMBENT LOCAL EXCHANGE CARRIERS (ILECS)

By and large, ILECs tend to be relatively homogeneous in terms of the products and services they provide. Their strengths lie in the access and transport business at which they excel, but because of the commoditization of this business, their ability to maintain marketshare is diminishing. For the most part, they have grown by acquiring more of the same. Witness Bell Atlantic's acquisitions of NYNEX and GTE, or SBC's acquisitions of Pacific Bell, Nevada Bell, Ameritech, and SNET. They have expanded their footprint but have not done much to diversify their product and service offerings.

In fairness, this strategy is used for good reason. The ILECs have realized that with long-distance relief pending, they must create a wide area presence for themselves. Their larger business customers are not necessarily local companies. Although they may have a local presence, they tend to be national or even global. If the ILECs are to become full service *providers*, they must be able to serve those customers on an end-to-end basis, thus eliminating the need for intermediaries. Without a wide area data network, they cannot accomplish this.

The fact is that the market is the ILECs' to lose. Many analysts believe that customers will buy all services from the local service provider if it has the ability to provision them. The holder of the access lines rules. Consequently, much of the company convergence activity of late has revolved around the acquisition of access lines. Consider Qwest's acquisition of USWest or Global Crossing's acquisition of Frontier. On a slightly different level, AT&T's acquisition of TCI is clearly a gambit for local loops and more will follow.

Today ILECs fall largely in the flea market model. The product they sell is fast becoming a commodity, and although they are good at what they do, they do not add a great deal of additional value above and beyond the access and transport services they provide. Although the flea market model tends to reflect more of a peer-to-peer management style, the ILECs exhibit more of a hierarchical management model.

So, are ILECs a dying breed? Will they be brought down by the smaller, more nimble CLECs that are nibbling away at their long-standing customer bases? There is no question that they face some serious challenges. Their networks were designed around the idea that they would control 100 percent of the market and are therefore not the most cost-effective resource in an open and competitive market. Other models are far more cost-effective than the ILECs' circuit-switched infrastructures. As a result, the ILECs are reinventing themselves a piece at a time and are, of course, expanding their market presence in a variety of ways.

THE COMPETITIVE LOCAL EXCHANGE CARRIER (CLECS)

The CLECs are similar to the ILECs in that they sell a commodity. They differ from the ILECs, however, because they can. They tend to sell the more lucrative products and avoid the markets and services that don't enjoy high returns. For example, many CLECs focus on residential voice customers, while others go after business customers. According to a study conducted by *Data Communications Magazine* in September 1998, out of approximately 500 CLECs (most of them in New York and California metropolitan areas), only 15 were going after major business customers at the time. Although the numbers have since increased, the disparity still exists to a degree. All CLECs are not created equal; their business strategies and business plans for carrying out those strategies and satisfying customers vary dramatically from company to company. They are, however, good performers within their identified market niches. The CLECs are obviously after the same access lines that the ILECs want to protect. The ongoing convergence of medium-size CLECs such as ICG, NextLink, and Intermedia are nothing more than positioning moves.

CLECs face significant obstacles by virtue of the fact that they are CLECs. As alternatives to the ILECs, they rely on interconnection agreements with them because they must have a collocation presence within the ILECs' central offices to provide service. The 1996 Telecommunications Reform Act mandates that before the ILECs will be allowed into the long-distance market, they must demonstrate that they have opened their local market to competitors, allowing equal access to unbundled facilities such as local loops and certain services. CLECs often complain that while the ILECs have agreed to the stipulations, they are not particularly quick to respond to CLEC requests for interconnection services and therefore have the ability to exert some control on the pace at which CLECs can enter their markets. Of course, some checks and balances are in place, such as Section 251 of the 1996 Communications Act. This component of the law requires that ILECs sell circuits, facilities, and services to their competitors that are "at least equal in quality to that provided by the local exchange carrier to itself or to any subsidiary, affiliate, or any other party to which the carrier provides interconnection."

CLECs are similar to the ILECs; they fall somewhere between the flea market and arbitrator business models, selling commodity access and transport while trying hard to add services that yield a competitive advantage over the ILECs. The ILECs control 90 percent of the access lines in the U.S., so CLECs face a significant challenge. Some of them, however, have proven to be innovative players. Teligent and Winstar, for example, offer wireless local loops that are targeted at small to midsize businesses, many of whom feel as if they do not get adequate attention from the ILECs. The service requires line of sight, but Winstar claims to be able to serve 75 percent of the buildings in a downtown area with the technology, which is LMDS-based. Furthermore, many CLECs claim to be able to offer better, more customized service than their ILEC competitors.

Most customers agree that the technology products sold by the ILECs and the CLECs are identical. The difference, they claim, is the way they deal with their customers. CLECs believe themselves to be more customer-focused, claiming that the ILECs are still plagued by legacy monopoly mentality. Whatever the case, some CLECs have initiated differentiation programs to help them garner the favor of customers such as payback plans for downtime, online real-time usage reports, and negotiated service-level agreements. Charlie Thomas, the president and CEO of Net2000, a CLEC based in Herndon, Virginia, believes that the key to success in the CLEC game is customer service. He left Bell Atlantic to form the company and believes that the ILECs will eventually lose as much as 40 percent of their current marketshare to competitors.

Ultimately, the success of the CLECs relies on three critical success factors. Local number portability must be viable, functional, and available; operations support standards must be in place and accepted; and discounts for unbundled network elements from ILECs must be on the order of 50 percent.

THE INTEREXCHANGE CARRIERS (IECS)

The IECs face the greatest challenge of all the players but are also the companies demonstrating the most innovative behavior in the face of adversity. With companies like Qwest and Level 3 building massively "overcapacitized" fiber networks, bandwidth is becoming so inexpensive and so universally available that it is evolving into a true commodity. The margins on it therefore are dropping rapidly. Furthermore, the number of companies that have entered the long-distance market has grown, as has their diversity. Although the legacy players (AT&T, Sprint, MCI) continue to hold the bulk of the market, a collection of power companies, satellite providers, and "bandwidth barons" have entered the game and are seizing significant pieces of market-share from the incumbents.

In response, the IECs are fighting back by diversifying. All have entered the ISP game, offering Internet access across their backbones at competitive prices. They have also bought or built local twisted-pair access infrastructures, cable companies, satellite companies, and a host of others. Consider AT&T as an example. Beginning with their long lines division, the company has grown into a multifaceted powerhouse that now owns IBM Global Networks, TCI, Teleport, Excite, and @Home, to name a few. They are a long-distance, local, wireless, ISP, portal cable company and have aggressive plans to be a true full-service telecommunications provider as they flesh out their strategy for market positioning.

The IECs, like the ILECs and CLECs, fall somewhere between the flea market and arbitrator models, although they are edging toward the arbitrator side of the equation as they diversify their holdings. Some of them, like AT&T, are becoming virtual corporations as they diversify.

THE INTERNET SERVICE PROVIDERS (ISPS) AND PORTAL PROVIDERS

The ISPs represent something of a mixed bag of business models. The small ISPs that provide nothing more than IP access to the Internet fall solidly into the flea market model, inasmuch as they provide a non-differentiated service among a host of other companies doing the same thing. The larger ISPs, however, have differentiated their service offerings to varying degrees and have formed alliances with other players in rather innovative ways. They have become portal providers as well, offering one-stop shopping/searching/gathering capabilities to their customers. AOL is the most visible example of a highly differentiated ISP. First of all, the company's service is targeted at the average consumer, not the "propeller-heads" that many ISPs cater to. Second, the company is diverse. It owns CompuServe Computer Online Services, Netscape, and ICQ and has strategic alliances with a number of key players including Sun Microsystems, Gateway, and TiVo. It has marketing agreements with SBC and Bell Atlantic, and hundreds of agreements with content providers to serve as their sole hosting service. To their credit, AOL has nearly 20 million customers and shows no signs of slowing down.

So where do the ISPs stand? The pure, low-end ISPs are clearly flea market companies, offering very little value integration. The larger ISPs, however, which host diverse content and provide portal services, fall more into the arbitrator model. They aggregate buyers and sellers, providing a safe, comfortable place for them to meet and do business. In some cases, they may exhibit characteristics of the process improver, creating order from chaos through business process enhancement.

THE CABLE COMPANIES

Similar to the ILECs, cable companies provide a largely homogeneous service that is under attack by a variety of alternative providers. Although they are striving to provide diverse services, the challenge to do so is daunting. TCI is in a good position through its alliance with AT&T and others will enjoy similar relationships in the future. In the meantime, however, they sell commodities and are therefore flea markets.

THE MANUFACTURERS

Faced with serious competition, rapidly advancing new technologies, and demands for service from the marketplace, the major (successful) equipment manufacturers exhibit company convergence at its finest. In the last three years, Cisco has acquired more than 40 companies, all designed to expand their abilities in the burgeoning telecommunications market. Long known as a data company, Cisco has reinvented itself and now offers an impressive line of integrated voice products as well. Nortel Networks and Lucent Technologies have followed similar routes, building products that will satisfy a diverse set of required abilities. As such, the manufacturers are flea market players, although they exhibit characteristics of all four company types.

EVERYBODY ELSE

So who falls into this category? Power companies occupy the largest niche, selling bandwidth on their own networks when the price is right. Other companies such as railroads and pipeline companies are also found here, but they are largely based on the flea market model.

AND THE POINT IS?

> Alice: Which way should I go?
> Cat: That depends where you are going.
> Alice: I don't know where I'm going!
> Cat: Then it doesn't matter which way you go.
>
> From Lewis Carroll's
> *Through the Looking Glass*

Three basic truths govern the decisions that companies are making today to define themselves within the model described in the last section. The first of these is that access and transport are fast becoming commodity products that can be had from a variety of providers at ever-lower rates. Second, if access and transport companies want to continue to maintain an edge in their marketplace, they must climb the service food chain and become more than what they currently are. They must become full-service telecommunications providers, because more and more customers want to buy all their needed services from a single provider. All services, incidentally, implies exactly that. The ability to offer a broad spectrum of bandwidth in multiple flavors is not enough. Customers want services, content, Internet access, and more, all delivered properly and billed under a single, itemized invoice.

The third of these is an observation about technological evolution. Wireless access is creeping in, and the percentage of fixed wireless local loops is climbing steadily. Some reports claim that as many as 40 percent of today's wire-based local loops will be replaced by wireless in the next few years. No single company in the telecommunications industry today is capable of accomplishing this on their own, yet their survival is based on their ability to do so. The only solution is to pool resources with other players, creating a "business molecule" that combines the best of multiple service "atoms" to create the combination of strengths and abilities desired by the customer base. Certainly, a need exists for all forms of business structure, as defined in the previous section, but it should be relatively clear at this point that service providers that have plans to stay in the game *must* take steps to move themselves into the virtual corporation quadrant. The reasons for this are clear. First, they recognize that they are simply not capable of providing the services customers desire today, given their current business model. Second, inasmuch as service is the name of the game, they must form alliances with other corporations so that they can provide a packaged bundle to their customers as a defense mechanism against competitors. Third, if they manage their mergers and alliances properly, customers will be unaware of the fact that in reality they are receiving services from many providers because they will be doing business with a single entity, behind which lies a cloud that disguises the complexity of the corporation providing service to the customer.

Five companies that elegantly demonstrate this converged mentality are Lucent Technologies, Cisco, AOL, AT&T, and Qwest. Lucent Technologies and Cisco are both telecommunications equipment manufacturers and powerful competitors. Lucent is well known for its flagship 5ESS circuit switch, while Cisco is recognized as the premier provider of packet switching-based products. AOL is an online service provider with plans to recreate itself as a virtual corporation, while AT&T is assuming the role in the service provider space. We will examine each of these companies in turn.

COMPANY CONVERGENCE: LUCENT TECHNOLOGIES

Lucent is a classical example of a virtual corporation. Originally founded as Western Electric, the manufacturing arm of the AT&T corporation, Lucent has undergone a number of significant changes over the years. Since being spun off from AT&T in 1996, the company has acquired more than 30 companies including

- Ascend for ATM core switching
- Yurie for ATM edge switching
- Optimay for GSM call-processing software
- Stratus for SS7 switching
- Quadritek for IP software
- Sybarus for semiconductor design
- WaveAccess for high-speed wireless Internet access
- Livingston for remote access products
- Prominet for Gigabit Ethernet
- Nexabit for IP switching and routing
- Mosaix for front-end/back-end system integration
- SpecTran Corporation for optical fiber
- *International Network Services* (INS) for consulting and systems integration
- Excel for IP-based programmable switching

If we consider the products represented here, we find that Lucent has done an exceptional job of shoring up the weaknesses in its product line, specifically the data side of the telecommunications house. The 5ESS is a well-known solution for telephony providers, and Lucent has long been recognized as a leader in legacy voice. It has, however, had a name as a data solutions provider until recently when its acquisition activities started shortly after being spun off of the AT&T juggernaut. Today the corporation has clearly become a multifaceted player with diverse capabilities that include the following:

- Circuit switch capability, including signaling
- ATM core and access switch technology
- Fast and Gigabit Ethernet
- Wireless telephony and Internet access
- IP platform software
- IP switches and routers
- Front-end to back-end system integration
- Systems consulting and integration

Clearly, this company has realized that it cannot continue to be a traditional provider of circuit-switched technology and has taken aggressive steps to correct their course. They have relied on the adage that "the quickest way to become a leader is to find a parade and get in front of it." The parade, it seems, is the long string of companies Lucent has either created alliances with or acquired. Lucent has reinvented itself as a virtual corporation capable of delivering every possible technology solution that a customer might request. In fact, the recent announcement of their 7 R/E switch platform, the evolutionary replacement for the 5E, reflects the company's acquisitions and alliances to date. The new switch boasts components and software from many of the companies that now bear the Lucent logo, yet when customers buy a 7 R/E, it bears the Lucent brand.

COMPANY CONVERGENCE: CISCO

Lucent, of course, is not alone. Cisco Corporation is another excellent example of the virtual corporation model. Founded in December of 1984 in Menlo Park, California, Cisco now has more than 10,000 employees working in 200 offices in 54 countries. Cisco does not rely on a single technology approach to the market. Their philosophy consists of listening to customer requests, assessing technological alternatives, and giving customers a range of options from which to choose. Their products are designed around internationally recognized standards, although some of their products rely on proprietary solutions designed to satisfy specific problems not adequately addressed by existing standards. In fact, some of Cisco's technologies have *become* industry standards. Cisco was also the first company to deliver multiprotocol routers. This gave them a head start over the competition and allowed them to become the leading supplier of enterprise networking solutions. To maintain market leadership, the company has rapidly developed and introduced new products and has effectively maintained its existing products. In response to their own interpretation of the marketplace, Cisco organized itself into six business groups: the Core Router group, the Access Router group, the Workgroup Products group, the ATM group, the IBM Market group, and the Internet Business group.

Additionally, Cisco has formed alliances with a number of WAN technology and service providers to develop flexible options for IP and frame relay services and has expressed a significant interest in the voice services marketplace.

Clearly, Cisco, as a router company, could not in and of itself address the needs of such a diverse market. To date, the company has acquired the following corporations, along with the capabilities each provides:

· Crescendo Communications for FDDI/CDDI hubs
· Lightstream for ATM switching equipment
· Grand Junction Networks for Ethernet and Fast Ethernet switches

- TGV for intranet and Internet software
- StrataCom for ATM and frame relay switches
- Nashoba for token ring switches
- Telesend for D4 DSL frame multiplexers
- Skystone systems for *Synchronous Optical Network* (SONET)/*Synchronous Digital Hierarchy* (SDH) equipment
- Global Internet software for security software in Windows NT systems
- Ardent Communications for compressed voice, LANs, and data over ATM and frame relay
- Fibex for digital loop carrier equipment that handles circuit-switched and ATM voice traffic
- Sentient for Gateway equipment between ATM and circuit-switched voice
- Amteva for unified messaging support software
- GeoTel Communications for support for distributed voice call centers
- Dagaz for XDSL equipment
- Light Speed International for voice signaling support
- Wheel Group Corporation for end-to-end network security software
- NetSpeed for DSL equipment
- Precept Software for video over IP
- CLASS Data Systems for priority resource control (*quality of service* [QoS])
- Summa Four for voice-over-IP switches
- American Internet Corporation for IP address management
- Clarity Wireless for fixed wireless
- Selsius for IP PBX
- PipeLines for SONET/SDH routers
- V-Bits for digital video over cable
- WebLine Communications for e-mail routing software

- Cerent for ATM frame relay and IP over fiber
- Monterey Networks for multi-gigabit optical routers
- Calista, Inc. for circuit and IP voice integration
- IBM's Networking Hardware Division for switching and routing

Additionally, Cisco has enjoyed strategic alliances with a dizzying array of companies, including its recent announcement of a $1.05 billion investment in systems integrator KPMG Consulting, as well as with AT&T, Hewlett-Packard, Intel, IBM, GTE, MCI, Microsoft, Ameritech, Sony, USWest, Fujitsu, Wang Laboratories, and many others.

If we analyze this remarkable collection of acquisitions and alliances, we discover that Cisco has become much more than a California-based router company. Ignoring the names of the companies for a moment, we find the following:

- ATM switches
- Frame relay switches
- Ethernet, Fast Ethernet, and token Ring switches
- DSL hardware
- SONET/SDH hardware and software support
- Network security
- Multiprotocol routing capabilities (including voice, video, LAN, and data)
- ATM-to-circuit-switched traffic gateway capabilities
- Telephony signaling (with IP to SS7 conversion)
- Unified messaging
- Call center support
- QoS support
- Wireless local loop
- IP and circuit-switched PBX capabilities
- Voice over IP switches

We are compelled to ask the following question: What is Cisco striving to become? Based on the list shown previously, and the business units they have rallied around (core routers, access routers, workgroup products, ATM, the IBM Market, and the Internet), they intend to be in the wide area core network business, as evidenced by their array of ATM and frame relay devices; the edge or access business, which their routers clearly satisfy; the Internet business, which their IP products clearly support; the IBM business, which their token ring products support; and the workgroup market, which is reflected in their LAN, QoS, unified messaging, call center, and network security product array. Obviously, Cisco intends be a full-service provider, and the company's alliances with integrators, standards bodies, content providers, local and long-distance telephone companies, consultancies, and other hardware providers reflects this.

COMPANY CONVERGENCE: AMERICA ONLINE (AOL)

The October 1999 issue of *Money Magazine* included an article entitled, "AOL: The One Stock You Can't Ignore." Tumult has certainly characterized this corporation, but in the long run, it has emerged as a shining success story among its peers, culminating with the acquisition of Time/Warner in January of 2000. The price, an almost unimaginable $154 billion, represents the largest merger in history, bigger even than the Sprint-MCI WorldCom mega-corporation. The benefits that accrue to both corporations are staggering. The combined corporation will be in a position to offer a complete suite of services to customers, including both media and information content. AOL's 20 million customers who rely on AOL for Web access will now also have access to Time/Warner's vast content holdings. It also provides AOL with a broadband cable network of its own for distribution of AOL content.

Within the dot-com stocks, AOL, barely 14 years old, is something of a silverback. It started in 1985 (the Internet equivalent of the late 16th century) as Quantum Communications Services, providing an interconnection service called Q-Link for users of Commodore 64 computers. Q-Link was followed in short order by AppleLink Personal Edition for Apple II users, PC-Link for Tandy Deskmate computers, and Promenade for IBM PS/2 users. In 1989, the name changed to America Online, and all services were discontinued so that the company could focus on their concept of a single product that would work for everyone. The initial service rolled out shortly thereafter for Apple II and Macintosh computers, and in 1993 a Windows version appeared.

"Buried in its success lie the seeds of its own destruction" is a phrase that characterizes AOL's rise to stardom rather nicely. In 1997, the company introduced a flat-rate pricing plan that, coupled with the enormous interest in the Internet and AOL's burgeoning customer base, contributed to severe telephone company central office switch congestion and the inability of users to access the AOL network due to insufficient modems in the company's modem pools. After numerous court actions and nationwide appeals for relief from customers, the company beefed up its network and today enjoys revenues from 20 million customers worldwide.

AOL realized early on that the number of ISPs was growing and that if they were to stay ahead of the competition curve they would have to offer more than simple access to the Internet. The company has a long history of crafting strategic alliances with hundreds of content providers who make their information available within the AOL domain so that subscribers do not have to leave AOL to reach it. Careful studies showed that a significant percentage of the company's subscribers never leave AOL; that is, they never actually use the button that enables them to go to the Internet itself, because they find everything they need within the system. Consequently, AOL decided to make themselves over, becoming a full-service provider offering much more than simple Internet access. They became a portal provider, an e-commerce leader, and a provider of interactive chat capabilities. Through a series of carefully planned acquisitions and mergers, AOL was able to expand into a diversity of service areas. In addition to Time/Warner, these acquisitions include

· Netscape for browser software
· When.com for Internet calendar and events software
· PersonaLogic for Interactive consumer buying guides
· Mirabilis for ICQ instant messaging and chat capabilities
· CompuServe—they sold the network, kept the customers!
· Gateway ($800 million) for computer hardware
· Digital Marketing Services for branded interactive services
· MovieFone for movie listings and ticket info
· MapQuest for online maps and travel information

The company has also formed strategic alliances with a number of companies including

· SBC Communications
· Bell Atlantic
· Wal-Mart
· Circuit City
· VitaminShoppe
· eBay
· CBS
· Metrocall
· Packard Bell
· Telstra
· 3Com
· ABC
· New York Times
· E*Trade
· BBN
· Excite

AOL has clearly found a variety of ways to position itself before its existing and prospective customers, and to take advantage of the capabilities of its strategic partners. By forming alliances with Bell Atlantic and SBC, AOL has provided good rationale to the ILECs for the deployment of DSL, which they have been slow to roll out in some cases. For $40 a month, a customer can have high-speed access to AOL services over the ILEC's DSL network.

Like Lucent and Cisco, the company exhibits a form of convergence best described as heterogeneous; that is, the company is engaging in mergers and acquisitions designed not only to expand the company's online service, but also to move AOL into a variety of new, related services that complement one another in rather innovative ways. AOL's vast footprint as an ISP gives it and its content providers access to an enormous worldwide market, and their diverse services attracts customers from all walks of life. A sampling of the services available from AOL includes access to nearly 100 major magazines; the Shopping Channel, which gives subscribers online access to more than 300 retailers; Starbuck's, which provides coffee-related information; Thrive, which provides health-related information; NetGrocer, which enables subscribers to order groceries online; Digital City, which provides local news and information about a variety of cities; and, of course, the services that come packaged as part of the Time/Warner deal. They have also made forays into long-distance telephone service and may soon offer network-based office automation applications such as word processing, spreadsheets, and presentation software. As *an Application Service Provider* (ASP), they would no doubt capture a healthy proportion of the overall market, but *not* as a single service ISP.

COMPANY CONVERGENCE: AT&T

As the largest long-distance provider in the U.S., AT&T is a household name held in high regard by many. Their strategy, however, underwent a dramatic shift a few years ago when it became clear that AT&T's goal was to acquire a large piece of the local access market in addition to its significant long-distance presence. To achieve this goal, the company acquired a collection of companies, including

- Tele-Communications, Inc. for cable access
- Teleport Communications Group for local voice and data access
- IBM Global Network for network services and outsourcing agreements

Additionally, the company has strategic alliances with

- BT
- Time/Warner cable
- USWest
- MediaOne Group
- Comcast
- Cox Communications
- Cablevision Systems
- @Home
- Cisco Systems
- General Instrument

AT&T also has holdings in many other companies throughout the world, giving them the ability to provide global voice and data sales and service. They clearly understand the three key factors described earlier. By expanding into cable, wireless, and traditional twisted-pair local access, they have the ability to offer local service across a variety of access technologies, including the explosive family of wireless options. Furthermore, because of their long-distance presence, they can provide the one-stop, full-service shopping that customers have begun to demand and, with their various Internet access holdings, also have a major presence in the ISP arena.

COMPANY CONVERGENCE: QWEST

Although Qwest bills itself as an Internet company, it is actually much more than that. Industry analysts scratched their heads in confusion when the company purchased USWest, but other actions on the part of the company were far easier to understand. The company has strategic alliances with such powerhouses as Oracle, Netscape, Siebel Systems, and SAP America, all of which will help them develop customer service applications and portal capabilities. Bellsouth owns 10 percent of Qwest, and as soon as Bellsouth is allowed to enter the long distance market, the company will serve as a local sales arm in the southeast for Qwest.

The acquisitions continue. Qwest bought $15 million of Covad and Rhythms, giving them local DSL access in more than 50 markets. They also acquired Icon CMT, now called Qwest Internet Solutions, and have formed joint ventures with Microsoft, Hewlett-Packard, KPMG, and KPN.

THE SPECIAL CASE OF THE ILECS

The ILECs face competition from a variety of sources and must therefore take steps to protect their besieged marketshare and customer base. A number of technical factors have made them understand the steps they must take to remain competitive. These factors include the following:

- Cable and wireless technology providers pose the greatest threats to the ILECs.
- Cable modems will dominate the market for high-speed Internet access, acquiring more than 50 percent of the marketshare by 2005.
- Broadband wireless services such as LMDS will become strong competitors in the next few years.
- As the ILECs lose marketshare, the cost to serve a shrinking customer base becomes prohibitive.

· Wireless and cable infrastructures are less expensive to
  implement than the traditional wire-based infrastructures
  deployed by the ILECs.

For the most part, the ILECs have exhibited a form of
"enhanced homogeneous company convergence," characterized
by acquisitions of other local providers designed to expand their
footprint because of the need for wide area data transport capabilities. SBC, for example, has acquired Pacific Bell, Nevada
Bell, Ameritech, and SNET, while Bell Atlantic, the country's
other powerhouse, has acquired NYNEX and GTE. These
acquisitions are clearly homogeneous in nature. They add nothing new to the acquiring company's capabilities other than geographical coverage.

Other activities, however, add to the competitive mix.
Recognizing the strategic importance of wireless as an alternative access scheme, the ILECs' expansion into wireless properties has been spectacular. In September of 1999, Bell Atlantic
agreed to merge its wireless holdings with those of the U.K.'s
Vodafone Airtouch conglomerate, forming the largest mobile
telephone company in the U.S. with 20 million subscribers and
a combined net worth of $70 to 80 billion. The combined com-
pany's network will compete favorably with AT&T's, the second
largest player with 11 million subscribers.

As part of the company's strategy to increase its wide area
broadband presence, Bell Atlantic invested $2.2 billion in
MetroMedia in October of 1999. The investment will give the
ILEC access to MetroMedia's fiber optic network in 50 major
U.S. markets and a number of international locations as well.

SBC has been equally busy, forming alliances with AOL,
DirecTV, and Prodigy. Their wireless service, offered through
Southwestern Bell Wireless, Pacific Bell Wireless, SNET
Wireless, Nevada Bell Wireless, Ameritech Cellular, and Cellular One provides substantial coverage for local access on a
nearly nationwide basis. Ameritech, meanwhile, has formed an
alliance with ITXC Corporation, an Internet telephony services
provider that will now terminate calls on the Ameritech network.

USWest, recently acquired by Qwest, now forms part of a powerful combination of high-bandwidth, long-distance capabilities and local access in a number of major markets including Denver, Phoenix, Minneapolis-St. Paul, Seattle, Portland, Salt Lake City, and Albuquerque. The combined company can now provide long-distance voice and data transport, one of the major tenets of success in the evolving telecommunications market.

THE MISSING LINK

So what are these companies missing, if anything? The ILECs offer both narrowband and broadband wired and wireless local services, the capability to sell long-distance services as soon as FCC restrictions have been cleared, Internet access, and, in some cases, strategic relationships with content providers. What they do not have for the most part is a relationship with a systems integrator or applications consultant. In the same spirit that caused Lucent to acquire INS, Cisco to acquire a piece of KPMG, Novell to take a stake in IT consulting firm Whittman-Hart, and Nortel to acquire Clarify (the second largest manufacturer of customer relationship management software for call center applications, help desks, and mobile users), the ILECs must round out their holdings by adding integration and consulting services to their physical network holdings. Being a large provider of access and transport, even as a sole provider, is not enough, particularly when faced with the fact that IXCs and other market segments are aggressively looking to capture a substantial piece of the local access market. The critical importance of the *services* component of the telecommunications business is evident here, and knowledge-driven companies will win the game. If the ILECs are to protect their marketshare from the IXC intruders, they must offer a set of services that goes well beyond those offered by the competitors. To do this, they must recognize and respond to three factors.

The ILECs must design a technology-based information and knowledge derivation infrastructure. Although the secret to success lies in the ability to quickly and efficiently provision services and solutions, communications technology provides the critical underpinnings that make the services and solutions possible. The ILECs must therefore ensure that their networks are based on the right technologies and that they are available throughout their operating regions. They must also build a knowledge management infrastructure and inculcate in all employees the importance of knowledge sharing. According to a report by the Conference Board[3] in New York City, "Getting people to share information may well turn out to be one of the key managerial issues of the next few decades."

The ILECs must be able to move quickly, to be flexible, and to adapt to change readily. The ILECs must adopt managerial changes that will enable them to make decisions quickly and efficiently. This undoubtedly means evolving to more of a peer-to-peer model than they currently exhibit and moving decision-making power as low as possible within the organization. It is also a further reflection of the importance of knowledge-driven management. With the proper management of customer relationships and appropriate data-mining techniques, ILECs can anticipate marketplace changes and be prepared before the customer asks for them.

The ILECs must learn to sell solutions, not technologies. Telephone companies have been driven by technology for so long that it is difficult to change the focus from technology-based selling to solution-based selling. It is an absolute necessity, however, if they are to remain in the game. As we have observed before, customers no longer care about the underlying technology. They do, however, care a great deal about whether the solution offered by the service provider will help them solve their business challenges and better position them in their own competitive markets, which means that the focus must be on the customer's business and application requirements.

---

[3]"Knowledge Hoarding is an Old Business Habit that is Tough to Break." *The Wall Street Journal*, October 14, 1999, page 1.

THE VIRTUAL CORPORATION: SUPPORTING TECHNOLOGIES

In order for the virtual corporation model to work properly, technologies must be utilized that create the illusion that the customer is doing business with a single company and that help the various enterprise components of the virtual corporation work as a single entity.

When corporations join forces, an enormous amount of infrastructure work must be done if the merger is to function. Not only must organizational issues be managed, but also daunting technical challenges must be overcome. For example, corporate computer systems must be integrated, network infrastructures must be harmonized, applications must be joined, differing IT practices must be reconciled, and intra-corporate communications must be assured. Furthermore, these exercises must be done as cost-effectively as possible, considering the close scrutiny that most mergers attract.

ENABLING THE VIRTUAL PRIVATE NETWORK (VPN)

Because the virtual corporation comprises distinct and often far-flung entities, and because it is common in modern corporations to support remote workers, technologies have evolved that support the critical strategic demand for interconnecting multiple offices at a reasonable cost and providing remote access to corporate information. The extremely competitive nature of the marketplace today requires that employees have immediate and efficient access to the most current information available and that they be able to work remotely as efficiently as if they were directly connected to the corporate network. The other advantage has to do with extending the cloud. By creating a capable corporate network, customers can be brought in electronically to ensure a timely exchange of information, thus enhancing customer relationship management activities. One answer to this challenge is the VPN. A VPN is a cost-effective alternative to a dedicated facility. Instead of paying the mileage charges associated with private line, a VPN replaces the dedicated circuit with the public Internet, the use of which has no distance component associated with it.

VPNs represent an alternative to traditional dial-up access. In a direct dial scenario, a remote worker dials directly into a corporate modem pool at headquarters, which means that the cost of the connection is dependent upon the distance the worker is from the headquarters. In most cases, the employee uses a toll-free number, but a cost is associated with this technique nonetheless.

VPN access is quite different. When a remote user accesses a corporate network using a VPN, they do not dial directly into the corporate network. Instead, they create an Internet connection by dialing into a local ISP and then use special protocols that create a "tunnel" through the fabric of the Internet that safely transports the information between the remote worker and the corporate network. Because the Internet is serving as the WAN, no distance-related cost is associated with this technique. The "free" Internet simply becomes an extension of the corporate network, providing access for remote workers. To protect the user and the company, secure protocols enable information to be passed safely between a remote employee and corporate computer systems. The difference between these two techniques is shown in Figure 2-5.

Design considerations must be taken into account when implementing a VPN or a direct-dial solution. The first thing that must be determined is whether a VPN solution is merited. If the bulk of the users who will be dialing into the corporate network are geographically close to headquarters, a dial solution may be perfectly acceptable because they will incur no distance-related charges as a result of their sessions. If, however, the users are more geographically dispersed, then an Internet or alternative IP solution may be in order to keep the access costs down. The IT staff must perform an audit of all users to create a profile that will indicate to them whether a VPN solution is required.

Direct Dial

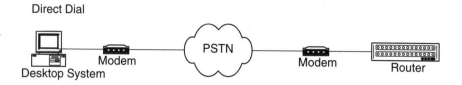

VPN

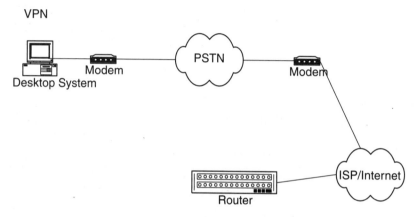

FIGURE 2-5   Direct-dial versus VPN connectivity

The second question that must be asked has to do with the number of ports that will be required on the headquarters server. For dial users, the IT staff must determine the number of modems that will be required, while VPN implementations are based on the number of simultaneous sessions that can be expected. Once the session count has been estimated, the total bandwidth required can be estimated based on the bandwidth required per session and the maximum number of sessions expected at any one time.

The third concern that must be addressed is network security. According to Gary Kessler, a specialist in network security, six questions must be answered when assessing security solutions:

- What are you trying to protect?
- Who are you trying to protect it from?
- What is the likelihood of an attack?
- What kind of attack is most likely?
- What would be the results of an attack?
- How much protection can you afford?

The answer to these questions comes in three parts: authentication, data integrity, and session privacy. Authentication is the process of determining that the user really is who they say they are and that the data being tunneled into the corporate network is from a reputable source. Because VPNs use the Internet as their transport fabric, it is critical to authenticate all users and the data that they send to corporate systems before they are allowed inside. This can be done in a number of ways, including simple login/password combinations, secure tokens, and digital certificates. Login/password combinations are the easiest form of security to break, because most users choose passwords that they can easily remember, which means that a hacker can easily determine what they are. One young hacker that I interviewed for an article on network security proudly boasted that "given five minutes with your wallet or purse, I'll have enough information to guess your passwords."

A more robust alternative to passwords is the secure token. Secure tokens are based on one of two factors: *something you have,* such as a secure ID card or one-time password, or *something you are,* such as a fingerprint scan, retina scan, or voiceprint. These techniques are becoming more common but still fall on the expensive side of the security spectrum.

Although passwords and secure tokens offer a modicum of security, digital certificates have emerged as the clear winner. The concept of a digital certificate is quite simple. It is nothing more than a "digital document" that contains a public key used to "read" secure documents, a name associated with that key, an expiration date, the name of the trusted third party that issued the certificate, a serial number, and information about appropriate user policies for the certificate. It may also contain the user's digital signature.

The certificate is used for the following purposes: to establish the identity of the user by associating a public key to an individual or organization, to assign authority by clearly stating the actions that the holder of the certificate may or may not take, and to protect the data that will be transmitted by encrypting the information using one of several secure protocols.

VPNs, then, represent a secure and cost-effective alternative to a dedicated network, making it possible to integrate far-flung corporate LANs and WANs using a public network infrastructure. Currently, private line and frame relay represent the bulk of all WAN installations. Unfortunately, they are relatively inflexible, a problem in today's dynamic network environment. And because they are usually sold under a three-to-five year contract, they have become less attractive for customers who do not want to be locked into a network solution that is not capable of evolving in lockstep with their changing applications and geographic footprint. Thus evolved the strong interest in VPNs.

The cost advantages of VPNs are quite significant. In addition to the savings that result from the elimination of private-line distance charges and dial access costs, reductions in capital equipment and support activities also contribute to fewer costs. Because VPNs use a single WAN interface for multiple functions, the data that would normally have passed through several data devices now requires one. Support costs are reduced because of the capability to consolidate all support functions within a single help desk organization.

Of course, VPNs offer challenges, not the least of which is performance. The Internet is not known for providing rock-solid QoS, and given that it constitutes the heart of the VPN, a number of questions arise. An ideal IP-based VPN would

- Be ubiquitously available, secure, and reliable
- Offer a variety of network management and billing options
- Provide measurable service level agreements
- Enable the user to differentiate QoS on a per-flow and/or per-application basis

The Internet, or even a pure IP network, does not inherently have the capability to accomplish all of these things. IP on top of an ATM backbone does, however, and this model seems to be the emerging choice for carriers looking to deploy QoS-capable public networks. By associating IP addresses to ATM virtual circuits, QoS can be assured.

So, what is the future of the VPN? First, it is important to acknowledge that the private-line network will never disappear, but as Internet QoS performance and encryption protocols improve, the need for a dedicated facility will be come less and less obvious. The VPN, because of its capability to reduce costs, support new business opportunities, and improve flexibility and speed to market will be recognized as a powerful enabler of customer relationship management and therefore competitive positioning.

THE EVOLVING COMPUTER

The evolution of the modern computer from the centralized mainframe to the fully distributed system has occurred in concert with the decentralization of the modern corporation and its growing focus on the need to reduce the distance between the company and the customer.

Contrary to what many believe, the mainframe computer is far from dead. Although it is no longer the center of the computing universe, it still plays a major role for corporations that need a centralized database and application support across the enterprise. Neither minicomputers nor networked PCs can provide this capability, so while its role has shifted, the mainframe still plays a central role in enterprise computing.

When IBM first rolled out its corporate mainframes several decades ago, they relied on a hierarchical approach to computing that emulated the managerial model of the corporations that installed them. *Systems Network Architecture* (SNA), released in 1974, followed the same model in which all roads led to the mainframe. Today SNA is thriving, but it is a very different beast than the centralized approach that first rolled out of IBM's doorway. Recognizing the evolution to a distributed computing model, IBM modified SNA to accommodate the needs of distributed corporations. *Advanced Program-to-Program Communication* (APPC) enables communication between peer-level programs without having to create a session through the mainframe, a first step away from the hierarchy of yore. *Advanced Peer-to-Peer Networking* (APPN) came shortly thereafter, enabling SNA to support the requirements of distributed computing. With APPN, minicomputers can be networked together to create a true peer-to-peer environment, moving IBM from its centralized roots into the departmental computing model while maintaining its position as the leading provider of mainframe technology.

A concern generating significant angst within IT departments these days is the issue of legacy integration. Although many corporations have significant investments in mainframe systems, there is tremendous pressure to integrate newer technologies, including IP. This represents another example of the adage that observes, "if it ain't broke, don't fix it." The fact that a corporation has legacy systems in place does not necessarily imply that they are outmoded. Many corporations have routinely maintained and upgraded their systems such that they are still high-performance machines. In fact, according to the Meta Group consultancy, most companies have found that it makes more sense to maintain legacy installations than to replace them with modern technology because 70 percent of all corporate data still resides on them.

However, pressure from TCP/IP is forcing IT managers to seriously consider merging it into their SNA-based systems. Good reasons exist for the inclusion of TCP/IP, including its ubiquity, its use in thin client systems, its flexibility, its error-recovery capabilities, and its enormous installed base, thanks to the popularity of the Internet. The greatest benefit, of course, is the universal nature of TCP/IP as a standard network protocol, and the fact that it is now included in every network operating system is an indication of its central role in modern networking.

TCP/IP was first introduced into the OS400 operating system, which enabled it to be easily integrated into the corporate computing environment. The challenge that had to be over-come in the newly integrated network was the need to route SNA traffic over a TCP/IP WAN because of the continued presence of legacy applications. A number of solutions have been proposed, most of which involve tunneling or encapsulation. These techniques involve the encapsulation of SNA data and include *Data Link Switching* (DLSw), which is the most common technique used. In DLSw, routers perform the actual encapsulation process, meaning that the end devices only need to handle SNA traffic with no need to understand TCP/IP. As a result, the integration of legacy environments with TCP/IP is relatively straightforward and is easily accomplished by competent IT staff.

Of course, the computing world has continued to evolve beyond the mainframe. The PC's impact has been immeasurably profound and, combined with high-speed LAN technology, enables the creation of massively distributed "virtual super-computers" throughout a corporation.

STORAGE AREA NETWORKS (SANS)

One of the principle advantages of a mainframe is the capability to manage and maintain a massive centralized database accessible by all users. One recent development is the *storage area network* (SAN). According to research firm *International Data Corporation* (IDC), SANs and the Fibre Channel[4] technology that interconnects them are fast becoming the preferred storage technology for enterprise data centers.

The problems that SANs are designed to correct are common. In distributed environments, data is stored on client

---

[4]Fibre Channel is an ANSI standard that defines a high-speed technique for transmitting data between all manner of data devices. It transmits at speeds ranging from 133 Mbps to one Gbps, and four-Gbps systems are under development. In spite of the name, Fibre Channel can operate over either fiber or coaxial cable. See www.fibrechannel.com for more information.

systems, managed servers, departmental servers, and mainframes. This results in a number of problems, including difficulties when accessing corporate information, uncoordinated database backups, lack of version control, the inability to manage and monitor performance, and uncontrolled storage costs. In a SAN, data storage is centrally managed to guarantee fault tolerance, scheduled backups of critical data, flexibility, and universal accessibility.

Although SANs offer a great deal to IT managers, problems still need to be worked out of this relatively new technology. First, interoperability issues must be resolved on both the client and server sides of the equation. Major vendors have announced Fibre Channel functionality, but some third parties do not offer compatibility with it. Another limitation is distance. Currently, Fibre Channel is limited to approximately six miles, although a 25- to 30-mile enhancement is in the works. As Figure 2-6 illustrates, user-attached LANs or other components of the corporate network can access any number of database components via the Fiber Channel hub, which provides universal, high-speed connectivity. A number of major vendors have announced SAN products, including Hewlett-Packard and IBM.

THE NEED FOR MOBILITY

Equally powerful has been the response of the computer industry to the need for mobility. The laptop computer is so common today that it is no longer a novelty, but the evolution doesn't stop there. The dramatic impact of palm-top computers, most notably the 3Com Palm Pilot, has created a whole new realm of computing capabilities. Today knowledge workers and "road warriors" often travel with nothing more than a palm-top and a cell phone with which they can accomplish most of the same tasks that formerly required an office, computer, and phone. Collaborative vendors have already released devices that combine the capabilities of a palm-top and cell phone, and further innovation is expected.

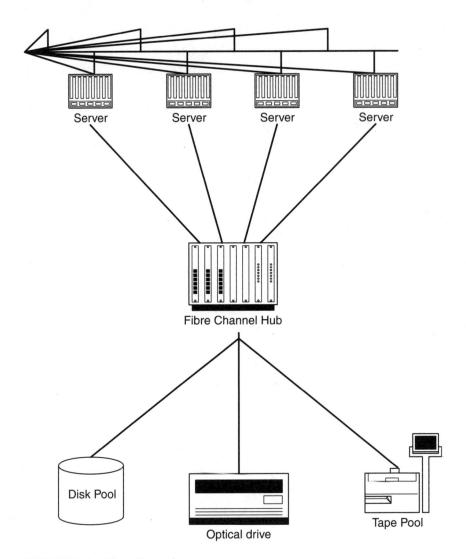

FIGURE 2-6   Fibre Channel

THINNING THE COMPUTER HERD

Another ongoing trend, although there is some confusion over whether "ongoing" means growing or shrinking, is the development of the so-called *network computer*, led by Larry Ellison of Oracle fame who first attempted to succeed with network computers in the mid-1990s using Oracle software and Sun Microsystems' Java[5] operating system. The idea behind a network computer is that the bulk of the intelligence required to make the machine useful resides in network servers rather than in the PC itself, thus making the client device inexpensive. From a corporate IT manager's point of view, network computers (sometimes called *thin clients*, another name from Oracle) represent the possibility of substantial savings because the PCs that are distributed to the corporation's many users are slimmed-down versions of a PC and are thus far less expensive. The intelligence resides in the servers that are shared among all users, making administration, maintenance, and therefore cost control significantly simpler.

SUMMARY

Driven by the need to provide a suite of services through consolidated technologies, company convergence has raged throughout the telecommunications industry. According to a report issued by Broadview International Investment Bank, mergers and acquisitions in the IT, telecommunications, and media sectors rose 18 percent in the first half of 1999 to more than 2,900 global transactions. In North America, M&A activity rose 47 percent to $351.4 billion, based on 1,805 transactions. As companies struggle to become the chosen single source provider for telecommunications services, the dealmaking accelerates as they form competitive clusters that can meet the challenges posed by customers in a marketplace that grows more competitive with every passing day. However, the clustering of companies is only a piece of the convergence process. They must also take whatever steps they can to ensure that they

---

[5]Java is a programming language invented by Sun Microsystems in 1995. It is designed to support distributed computing environments where applications reside on centralized Web servers. The applications can be downloaded from the Internet to a PC where they execute locally, thus eliminating the unpredictable latency of the Internet in the middle. Java is therefore ideal for distributed computing.

are functionally converged as much as they are organizationally converged. Otherwise, the customer will not gain any benefits.

This chapter focused on the evolution of the electronic business, an inevitable evolution in the Internet-driven business model. However, corporations must be aware of certain pitfalls as they progress to become fully functional e-businesses. According to Gartner Group, 75 percent of all e-business conversion projects will fail due to poor business planning and a lack of understanding of the technology required to bring about the evolution. The five most common reasons for failure include poor project management, a lack of clearly defined goals, inflexibility, minimal knowledge of competitors with regard to their electronic business activities, and a sense that the conversion to e-business is an end unto itself rather than a means to an end.

The traditional business model is changing, replacing wholesalers and retailers with online intermediaries and Web merchants. Those intermediaries are becoming specialized, focusing on specific industries or business processes that they can positively affect. Their services include service hosting, transaction mediation, the aggregation of buyers and sellers, and the provisioning of "safe harbors" for buyers and sellers. They serve as hubs, and their value is based on Metcalfe's Law, described earlier. Clearly, the more intermediary hubs that exist, the greater the aggregate value of the network and its services. These hub services generate significant revenue. According to investment bank Volpe Brown Whelan, hubs generated $290 million in 1998 and will grow to an almost inconceivable $20 billion by 2002.

The main challenge that these new service providers will face is the need to diversify into specific market segments. Thus, they have a choice. They can diversify vertically, which simply means that they will become industry-specific, or they can diversify horizontally, which implies that they will focus on business processes and work to make them more efficient, regardless of the industry. If a company chooses to diversify vertically, their success will increase depending on a number of key factors including the fragmentation between buyers and sellers, the ability to provide sophisticated online search capabilities, discrete knowledge of the industry they serve, the relationships that exist between the various players, supply-chain efficiencies, and the ability to aggregate major suppliers with buyers.

Similarly, horizontal diversification succeeds as a function of process standardization, discrete understanding of the business process of the industry served, knowledge of industry workflow, knowledge of process automation, and the ability to customize services and products to serve specific industry niches.

In the next chapter, we look at the final component of the convergence phenomenon, the convergence of services. There we reach the climax of the convergence process and pull it all together.

BIBLIOGRAPHY

BOOKS

Davis, Stan and Christopher Meyer. *Blue: The Speed of Change in the Connected Economy*. Addison-Wesley: Reading, MA, 1998.

Denning, Peter and Robert Metcalfe. *Beyond Calculation: The Next Fifty Years of Computing*. New York; Springer-Verlag, 1997.

Evans, Philip and Thomas Wurster. *Blown to Bits: How the New Economics of Information Transforms Strategy*. Harvard Business School Press: Boston, 2000.

Martin, James. *SNA: IBM's Networking Solution*. New York; Prentice-Hall, 1987.

Minoli, Daniel. *Telecommunications Technology Handbook*. Norwood, MA; Artech House, 1991.

Tapscott, Donald. *Growing up Digital: The Rise of the Net Generation*. McGraw-Hill; New York, 1998.

Tapscott, Donald. *Blueprint for the Digital Economy*. New York; McGraw-Hill, 1998.

WEB RESOURCES

"Achieving Business Success through Customer Relationship Management (CRM)." The Technology Guide Series, available at www.techguide.com.

"Business-to-Business E-Commerce: How Businesses Can Reduce Costs and Cash in on the Net Economy." A white paper available at www.ariba.com.

"The Challenge of Managing Knowledge." *Financial Times.* October 4, 1999.

"Cisco Releases Study Measuring Jobs and Revenues Tied to the Internet Economy." Company press release. June 10, 1999.

Cobleigh, Ward. "ERP on the Wire." A white paper available at www.ganymede.com.

"The Critical Role of Business Intelligence in E-Business." A white paper available at www.informationbuilders.com.

"Data Warehousing Today: Summaries and Subsets Can't Tell the Whole Story." The Technology Guide Series, available at www.techguide.com.

Garvey, Martin. "Legacy Systems: Reinvest or Restructure?" *Information Week Online.* August 9, 1999.

"The Internet Economy Indicators." www.internetindicators.com/features.html. June 11, 1999.

"The Internet Ecosystem: The Business model for the Internet Economy." www.internetindicators.com/ecosystem.html. June 11, 1999.

Isenberg, David. "The Rise of the Stupid Network." AT&T Bell Laboratories, 1997. www.camworld.com/stupid.html.

Karvé, Anita. "Lesson 136: Storage Area Networks." *Network Magazine.com.* November, 1999. Available at www.networkmagazine.com/magazine/tutorial/hardware/9911.htm.

"Knowledge Management Technology Solutions." The Technology Guide Series, available at www.techguide.com.

"PeopleSoft Enterprise Performance Management." An Executive Overview available at www.peoplesoft.com. February, 1999.

"Precise Sharpens E-business Performance Tools." *ZDNet*. October 4, 1999.

Simmons, Larry. "Extending Legacy to the Internet." A white paper available at `www.merant.com/ads/dc/wpaper/extending_legacy.asp`.

"Virtual Private Networking: Maximizing Network Performance While Reducing Costs." The Technology Guide Series, available at `www.techguide.com`.

ARTICLES

3Com Corporation. "In Phase With the Future." *Net Age*. Q3, 1999, page 5.

Alster, Norm. "The New AT&T: Now for the Hard Part." *Upside*. September 1999, page 120.

Borrus, Amy. "Why MCI's Brat Pack is all Over the Beltway." *Business Week*. August 30, 1999, page 172.

Burrows, Peter. "A Bright Ray of Sun at AOL." *Business Week*. September 27, 1999, page 58.

Cox, John. "How to Pick a Java Application Server." *Network World*. September 13, 1999, page 64.

Cross, Kim. "B-to-B, By the Numbers." *Business 2.0*. September 1999, page 109.

Dertouzos, Michael. "The Future of Computing." *Scientific American*. August 1999.

DeVeaux, Paul. "Nortel's Anytime, Anywhere Internet Access." *America's Network*. July 15, 1999, page 10.

Drucker, Peter. "Beyond the Information Revolution." *The Atlantic Monthly*. October 1999, page 47.

Duffy, Jim. "3Com Captain Remains Calm Despite Stormy Forecasts." *Network World*. July 5, 1999, page 55.

———. "Cisco Buys Voice Switch Maker Summa Four." *Network World*. August 3, 1998, page 11.

Dyrness, Christina. "Sayonara, Cisco." *TECHCapital*. September/October 1999, page 38.

Elstrom, Peter and Linda Himelstein. "What will AT&T do with Excite@Home?" *Business Week*. August 30, 1999, page 44.

Fontana, John. "Microsoft Shapes its Collaboration Platform." *Network World*. September 27, 1999, page 16.

Fox, Loren. "Another Face for Venture Capitalism?" *Upside*. October 1999, page 127.

Fried, John. "Cashing in, Moving on." *Washington Business Forward*. October 1999, page 24.

Garner, Rochelle. "Supply-Chain Leader Ventures into Internet-Based ERP." *Upside*. September 1999, page 53.

Greene, Tim. "AT&T, BT to Craft International Venture." *Network World*. August 3, 1999, page 23.

———. "The Vaunted VPN." *Network World*. September 27, 1999, page 65.

Gross, Neil. "21 Ideas for the 21st Century." *Business Week*. August 30, 1999, page 132.

Hammond, Eric. "Managing the Data Mountain." Special Report to Info World. October 11, 1999, page S1.

Hecht, Howard. "At What Price Ubiquity?" *TechWeb*. www.teledotcom.com/analysts_alley/analyst.html. July 22, 1999.

Helyar, John. "The One Stock You Can't Ignore." *Money*. October 1999, page 106.

Hotch, Ripley. "Check Mate." *Communications News*. April 1999, page 10.

"Is Sun/AOL Alliance a Mixed Blessing?" *America's Network*. May 1, 1999, page 13.

Keegan, Paul. "Can Bob Pittman, of MTV Fame, Make AOL Rock?" *Upside*. November 1999, page 98.

King, Rachel. "The New Breed of Telecom Companies." *TECHCapital*. September/October 1999, page 47.

Kovac, Ron. "VPN Basics." *Communications News*. April 1999, page 14.

Levine, Shira. "The ABCs of ERP." *America's Network*. September 1, 1999, page 54.

Lewis, Jamie. "As E-Business Evolves, Boundaries Will Give Way to Virtual Enterprise Networks." *Internet Week*. September 13, 1999, page 25.

Lippis, Nick. "Lucent vs. Cisco: The Coming Showdown." *Data Communications.* September 1998, page 19.

MacLeod, Alan. "Protecting the 'Private' in VPN." *Network World.* September 27, 1999, page 33.

Makris, Joanna. "Ticket to Hide?" *Data Communications.* September 1999, page 49.

Messmer, Ellen. "Financial Firls Investing in Web-Based Customer Service." *Network World.* October 11, 1999, page 43.

"Mindspring Loses its Bounce." *TECHCapital.* September/October 1999, page 15.

Moeller, Michael. "Maybe This is One Race Microsoft Can't Win." *Business Week.* March 22, 1999, page 37.

Nolle, Tom. "Qwest and USWest Have Their Roles Reversed." *Network World.* July 19, 1999, page 51.

Pappalardo, Denise. "AT&T WorldNet Still Mending Dial-up Net." *Network World.* July 5, 1999, page 21.

Pfeiffer, Marc. "Capitalizing on IP VPNs." *America's Network.* May 15, 1999, page 40.

Port, Otis. "Machines Will be Smarter Than We Are." *Business Week.* August 30, 1999, page 117.

"Put Some Backbone Into Your Decisions." *Communications News.* April 1999, page 22.

Rohde, David. "AT&T Wins Cable Battle, but War with RBOCs Continues." *Network World.* August 2, 1999, page 12.

Sawhney, Mohanbir and Steven Kaplan. "Let's Get Vertical." *Business 2.0.* September 1999, page 85.

Schlender, Brent. "The *Real* Road Ahead." *Fortune.* October 25, 1999, page 138.

Sloan, Allan. "How We'd Make Our Millions." *Newsweek.* June 21, 1999, page 55.

Songini, Marc. "IBM Net Gear Set to go to Cisco." *Network World.* September 6, 1999, page 1.

Songini, Marc. Packeteer Brings SNA reliability to IP. *Network World.* September 6, 1999, page 21.

Southwick, Karen. "Java Takes Off." Excerpt from *High Noon: The Inside Story of Scott McNealy and the Rise of Sun Microsystems. Upside.* October, 1999, page 153.

Spears, Mit. "The Empire Strikes Back." *Upside*. September 1999, page 150.

Tehrani, Rich. "Nortel Weighs in on Internet Telephony." *Internet Telephony*. July 1999, page 6.

Willey, Richard. "How Cryptic!" *Communications News*. April 1999, page 18.

SERVICES
CONVERGENCE

In previous sections, we described the overall convergence process. Initially, the collection and analysis of customer interaction activity lead to an understanding of their current and future service requirements. Those requirements translate into an understanding of the technology necessary to satisfy those service requirements, but because service providers do not typically have all of the needed technology and know-how, they acquire it through mergers, acquisitions, or strategic corporate alliances. Through those activities they form virtual corporations that, if built correctly, have the ability to not only deliver, but to anticipate future customer service requirements. In a market as competitive as the telecommunications industry, the ability to differentiate products that are for all intents and purposes commodities becomes a matter of survival.

In this chapter, we delve into the end game, the actual services that customers are looking for and the technologies required to deliver them.

THE SERVICES

I've looked at clouds from both sides now,

from up and down, and still somehow

it's cloud's illusions I recall.

I really don't know clouds at all.

Joni Mitchell, *Both Sides Now*

It is important to remember that the service provider's view of services differs substantially from that of the customer. To be innovative and competitive, the service provider must view the services world through the eyes of its customers. Otherwise, it becomes far too easy to think of service and technology as synonyms for each other, which they most definitely are not. As we have said before, technology is a facilitator of service and, as Joni Mitchell's song suggests, should be completely invisible to the customer. All they should see is an opaque cloud from which are delivered the capabilities they require to be competitive within their own markets.

From the customer's perspective, services include the following:

- Voice
- Distributive video (television or *Video-on-demand* [VOD])
- Interactive video
- Image transport
- Narrowband and broadband data services
- Internet access
- E-mail
- Fax

Ideally, a service provider delivers them over a single or a limited number of network interfaces, guarantees that the quality of the delivered service will be in accordance with a service-level agreement that both parties have signed, and is always one step ahead of the customer in terms of providing for their telecommunications needs.

From the point of view of the service provider, several key components must be addressed if they are to meet their customers' demands, the most important of which is *quality of service* (QoS). To a customer, QoS means many things, including

- An intelligible, recognizable voice
- A wide selection of VOD movies
- High-quality videoconferencing
- Medical-image-quality transport
- A variety of data services, ranging from low to high bandwidth
- Anywhere, anytime Internet access
- E-mail
- Low-cost fax (possibly IP-based)

To a service provider, the concept of service takes on different meanings because he or she must be concerned with the technology that will be required to provide all of the capabilities described earlier. Their key concern is to build a network infrastructure that has the capacity, flexibility, service granularity, "growability," and robustness to meet customer requirements. In their terms, QoS may mean

- Adequate, easy-to-provision bandwidth
- Minimal delays
- Minimal jitters
- Scalability of network resources
- Simple interworking with other networks
- Minimal information loss
- Enhanced management capability
- Security
- Reliability

As we discussed in an earlier part of the book, a significant interest exists in migrating to an IP network infrastructure because of the universality and flexibility that it enables. However, we also observed that although IP has significant advantages for service providers and customers alike, it does have some downsides associated with it, not the least of which is its incapability to provide QoS. Consequently, we now see service providers building multipurpose *asynchronous transfer mode* (ATM) core networks over which they run IP at the network layer. This hybrid approach has a number of advantages. First, it takes advantage of ATM's capability to provide granular QoS, seamless integration with *Synchronous Optical Networks* (SONETs)/*Synchronous Digital Hierarchies* (SDHs), and well-defined interfaces with a variety of access technologies. Its widespread international acceptance is also an advantage for global corporations.

Second, this hybrid approach enables IP's many advantages to be integrated into the mix. IP offers a universally accepted addressing scheme and the capability to interoperate with a broad range of widely deployed protocols. That, combined with the bandwidth and QoS of ATM, makes for a powerful and enormously flexible technology combination. If a service provider were to deploy ATM and IP as their core multiservice network, they would be able to offer access and transport services for any imaginable application over a single network infrastructure, thus realizing their own requirements for achieving higher service quality at a lower cost.

An example of QoS delivered over ATM can be seen with Ascend's IP Navigator product. Using a service called Absolute QoS, Navigator searches for IP packets with specific QoS indicators. It then attempts to associate the QoS requirement of the packet with a pre-established *switched virtual circuit* (SVC) between the sender and receiver. If it finds one, it immediately transmits the packets across the circuit. If it fails to make an association, it creates one with the appropriate characteristics necessary to deliver the QoS required.

All major vendors have announced products designed to facilitate interoperability between the voice and IP data worlds. In August 1999, Cisco announced a group of products designed to bring voice capabilities to their line of routers, including T1/E1 cards, and a multiservice router. They also announced software augmentations to their existing products, including the MCM H.323 Gatekeeper, a feature set in the IOS software that provides QoS, security, and service management for voice and video applications.

In February 1999, Argon Networks released their Giga-Packet Node, a high-speed router that offers native IP routing capabilities as well as native-mode ATM in the same device. Designed for enterprise-wide *virtual private network* (VPN) implementation, the GigaPacket Node reduces operation costs and supports a variety of QoS protocols including *Multiprotocol Label Switching* (MPLS). By providing ATM at the core, the device enables service providers to support a variety of payload types including ATM, frame relay, private line, and SNA, to name a few.

VIEWING THE NETWORK: INSIDE OUT OR OUTSIDE IN?

The network can be viewed in one of two ways: from the perspective of the service provider or from the point of view of the customer. The service provider sees a group of access technologies that will connect to their core network, while the customer sees a transport cloud from which their access technologies will derive services. The network, then, comprises two distinct areas of responsibility: the *edge* and the *core*. The edge of the network is responsible for aggregating traffic, prioritizing it, and packaging it for hand-off either from the customer to the cloud or from the cloud to the customer. The core is responsible for maintaining the QoS requested by the customer's application on an end-to-end basis by managing the nine QoS factors mentioned earlier. We will discuss each of these in turn in this section.

Bandwidth is exactly that, a measure of the volume of information that can be transmitted through a channel in a given period of time, usually a second. A *delay* is the amount of time it takes for a packet to transit a network. As long as the delay is constant (that is, all packets are delayed the same amount of time), problems are minimized. When the end-to-end delay begins to be undependable, however, problems become significantly worse. This variability in transit time is called a *jitter*, which measures the variability of the delay from packet to packet between two endpoints.

Information loss is an indirect measure of the quality of the overall network. If packets are being lost on a regular basis, service quality suffers and delays may increase due to the need to resend the missing packets. *Scalability* is an indication of the ease with which the network accommodates growth. *Interworking* reflects the degree to which the network is designed around international standards and is therefore interoperable with other networks.

Network management is critical from the point of view of the service provider, because without network management, the network cannot be properly monitored to determine whether it is performing up to the requirements of the service-level agreement.

Security is an important selling point in any network, particularly when VPNs are to be implemented across a public network infrastructure. Finally, *reliability* is the principal factor when negotiating service-level agreements. To the customer's way of thinking, all of the factors listed earlier fall into this category. From the service provider's point of view, however, reliability is a measure of the network's technical performance.

The bottom line, then, is that customers have certain preconceived expectations about what they should receive from their service provider. As long as they understand those expectations, the service provider will be able to anticipate those requirements and build a network infrastructure that can handle them. Today IP is emerging as a major component of the so-called next-generation network for many good reasons that include economics, diversity, ubiquity, scalability, and capability. IP telephony is a rallying cry for IP's widespread deployment. In the next section, we examine how it works, how it differs from traditional telephony infrastructures, and how it will be implemented as a viable contender for voice service transport.

THE EVOLUTION OF IP TELEPHONY

IP telephony promises to reduce communication costs and simplify multiservice corporate networks because of its packet-based underpinnings. However, to fully understand IP telephony and all that it and other IP-based services have to offer to the service provider as well as to the customer, it is first important to understand the existing telecommunications network that it may someday replace.

When a call is placed via a traditional telephone network, the call connection process is simple and straightforward. The caller notifies the local switch of his intent to place a call by lifting the handset, which completes a circuit and enables a current to flow from the switch to the telephone.[1] The network responds to the caller's request for service by sending a dial tone from the switch to the telephone. The caller then notifies the switch of the destination address by entering the number to be dialed. The switch responds that it has received the appropriate number of digits by putting a ringing tone, busy signal, or fast busy signal in the caller's ear as appropriate.

Of course, there is much more to it than this going on behind the scenes. When the switch receives the dialed digits, it must first determine whether the call is local or long-distance by consulting a table that identifies local NXXs (prefixes). In the number 508-555-1798, 555 is the NXX. If the call is local and served by the same switch, the switch connects the call directly. If the called party is served in another switch, whether local or long-distance, the originating switch must reserve a trunk to connect the originating and destination switch and then send the call to the destination switch. If the destination switch is in a different geographic area, the call must first be handed off to a long-distance company, which will in turn hand the call back to the receiver's local service provider upon call completion.

---

[1]We will assume that this is an analog local loop and ignore ISDN for the moment. The concept, however, is the same. Instead of dial tones and other analog constructs, ISDN uses a variety of packets to control the call setup process.

The typical telephone network comprises a hierarchy of local and toll switches that provide a local interconnection between customers and the service provider, and ultimately between the customer and their chosen *interexchange carrier's* (IXC) network. The switches in turn are interconnected with an interstitial fabric of optical multiplexers, digital cross-connect devices, SONET or SDH add/drop multiplexing equipment, and a variety of other components. The entire system is controlled by a fully separate signaling network known as *Signaling System 7* (SS7), which is the protocol suite responsible for setting up, maintaining, and tearing down a call while at the same time providing access to enhanced services such as *Custom Local Area Signaling Services* (CLASS), 800 number portability, *Local Number Portability* (LNP), *Line Information Database* (LIDB) database lookups for credit card verification, and other enhanced features.

The original concept behind SS7 was to separate the actual calls on the *Public Switched Telephone Network* (PSTN) from the process of setting up and tearing down those calls in order to make the network more efficient. This had the effect of moving the intelligence out of the PSTN and into a separate network where it could be somewhat centralized and therefore made available to a much broader population. The SS7 network, shown in Figure 5-1, consists of packet switches (*signal transfer points* [STPs]) and intelligent database engines (*service control points* [SCPs]), interconnected to each other and to the actual telephone company switches (*service switching points* [SSPs]) via high-speed digital links, typically operating at 56 to 64 Kbps.

When a customer in an SS7 environment places a call, the following process takes place. The local switching infrastructure issues a software interrupt via the SSP so that the called and calling party information can be handed off to the SS7 network, specifically an STP. The STP in turn routes the information to an associated SCP, which performs a database lookup to determine whether any special call-handling instructions apply. For example, if the calling party has chosen to block the delivery of caller ID information, the SCP query will return that fact.

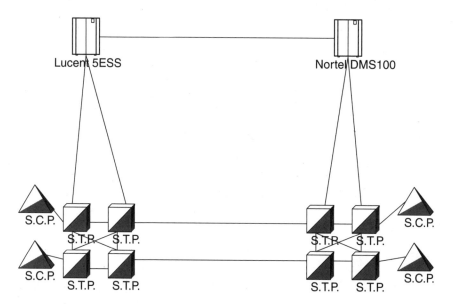

FIGURE 3-1   The SS7 network

Once the SCP has performed its task, the call information is returned to the STP packet switch, which consults routing tables and then routes the call. Upon receipt of the call, the destination switch determines whether the called party's phone is available, and if it is, it will ring the phone. If the customer's number is not available due to a busy condition or some other event, a packet will be returned to the source indicating this fact and SS7 will instruct the originating SSP to sound a busy tone or a reorder.

At this point, the calling party has several options, one of which is to invoke one of the many CLASS services such as *Automatic Ringback*. With Automatic Ringback, the network monitors the called number for a period of time, waiting for the line to become available. As soon as it is, the call will be cut through, and the calling party will be notified of the incoming call via some kind of distinctive ringing.

Thus, when a call is placed to a distant switch, the calling information is passed to SS7, which uses the caller's number, the called number, and SCP database information to choose a route for the call. It then determines whether any special call-handling requirements should be invoked, such as CLASS services, and instructs the various switches along the way to process the call appropriately.

These features comprise a set of services known as the *Advanced Intelligent Network* (AIN), a term coined by Telcordia (formerly Bellcore). The SSPs (switches) are responsible for basic calling, while the SCPs manage the enhanced services that ride atop the calls. The SS7 network, then, is responsible for the signaling required to establish and tear down calls and to invoke supplementary or enhanced services.

One problem with Telcordia's concept of the AIN is the fact that it is somewhat incomplete. It was designed to be an open standard and therefore vendor-independent. Unfortunately, by the time AIN arrived, a substantial embedded base of legacy equipment was already established. Thus, to implement AIN, equipment manufacturers had to modify their in-place hardware. This resulted in significant expenses and posed a serious obstacle to service providers wanting to deploy new services. In addition to that, the network management capability of the AIN was incomplete. The ability for service providers to manage existing services or to create new ones was difficult, and this factor slowed deployment as well. For these reasons, IP telephony has caught the attention of service providers who must move quickly if they are to be competitive in a marketplace characterized by new, low-cost, and aggressive entrants who do not have the burden of a legacy architecture.

In the IP environment, the goal is identical, but a different process is used. SS7 will still be used in the IP world, but the SS7 packets will be converted to IP-based signaling packets by some form of intelligent peripheral. A number of efforts are underway to directly interface IP to SS7, including Nortel Networks' IPS7 and similar protocols from other vendors.

With that understanding in mind, let's now consider how a long-distance IP telephone call might be processed using an *Internet telephony service provider* (ITSP). In Internet telephony, we must introduce a number of new devices and eliminate some old ones (see Figure 5-2). Because Internet telephony will exist in

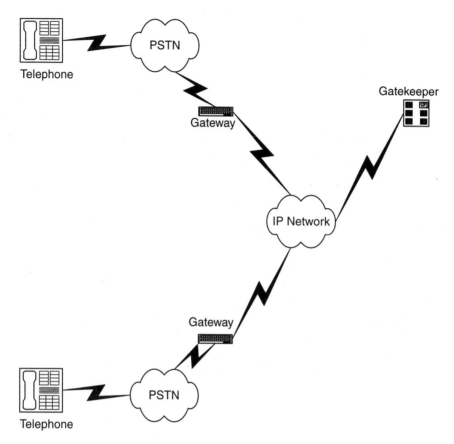

FIGURE 3-2   Interconnecting the PSTN and IP networks

parallel with traditional circuit-switched telephony for many years to come, local and long-distance switches will remain a reality in the circuit-switched domain. In IP telephony, however, we now introduce *gateways* and *gatekeepers*. A gateway is a device that performs destination address resolution, connection monitoring and management, circuit-to-packet conversion, voice digitization and compression, circuit-to-packet signaling, security, and accounting. Some gateways are robust enough to replace a local switch and all of them replace the functions provided by a toll switch. A gatekeeper, on the other hand, serves as the administrative authority in IP voice environments, managing such functions as bandwidth management, call zone management, address translation, and usage or admission control. Call zone management simply means that the network can be logically managed such that certain calls are restricted or otherwise controlled to satisfy the demands of management.[2] Address translation refers to the fact that a user can place a call to another user by entering their e-mail address. The gatekeeper then translates the address into a telephony address that the network can handle. Finally, admission control is simply a security/management function that enables calls to be managed based on the time of day, user authentication, address pooling, or other methods.

The most desirable benefit of IP telephony is its capability to use the Internet or a higher-quality IP network as a fabric for bypassing the traditionally expensive long-distance network. Thus, the call might proceed as shown in Figure 5-3. As in a traditional circuit switched call, the caller goes off-hook and dials a long-distance number. When the local switch receives the dialed digits, it recognizes that the call is not local and proceeds to hand it off to the customer's chosen long-distance carrier. In

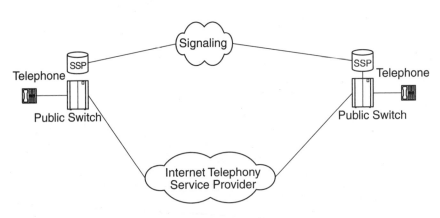

FIGURE 3-3   IP telephony

---

[2] A number of protocols are often used to control these functions and to establish the "zones" within an IP telephony network. H.323 is currently the most common, although the *Session Initialization Protocol* (SIP), the *Media Gateway Control Protocol* (MGCP), and their "relatives," discussed a bit later, are gaining popularity.

this case, the customer's long-distance provider is an ITSP, so the circuit-switched call is handed off to it. The ITSP converts the stream of information to a stream of IP packets that it must now route to the requested destination, all the while being cognizant of the fact that this is a voice call and must therefore be afforded strictly controlled QoS.

The ITSP must first resolve the address so that it can route the call to the proper gateway that interfaces with the local service provider that serves the destination customer. Once it has determined the address, it must route the call, making sure low delay is of paramount importance. At the same time, the ITSP must signal to the far end to ensure that the call can be connected through the called party's local service provider. The overall goal is identical; it is simply handled in different ways.

The greatest difference between circuit-switched voice and IP voice is the issue of quality control. In circuit switching, QoS is largely guaranteed because of the nature of the infrastructure. In IP voice, however, you have no guarantees, again because of the nature of the connectionless IP infrastructure. Thus, something must be done to ensure quality control if voice is to be transported over an IP infrastructure with anything resembling "carrier-class" service. One solution, of course, is to rely on a high-quality transport infrastructure such as ATM, which many companies are now doing. The alternative is to use emerging QoS protocols and devices capable of responding to their mandates. Those protocols and the devices that implement them are the subject of the next section.

EMERGING QoS PROTOCOLS

A conflict exists today in the area of QoS protocols, because they tend to be either Internet-driven or telephony-driven. Not until recently have there been efforts to reconcile the differences between the two and to move toward a single, common foundation of standards that serve the requirements of both the circuit-switched and packet-switched camps.

For a service provider to guarantee QoS for the purposes of service-level agreements, it must be able to manage both call signaling and content queuing. The signaling is required not only for the setup and teardown of calls, but also for the QoS requests that are an inherent part of signaling in a connectionless, IP-based network. Furthermore, every device along the flow of the data must be able to acknowledge and support the guarantee of delay and bandwidth. If a device is incapable of supporting the requested quality level, the call is refused.

To guarantee that delays will be minimized, switches and routers must be able to buffer flow state information, that is, the characteristics of each flow of information headed into or out of the network. Without that capability, the network cannot guarantee minimal delays, which means that delay-sensitive services such as voice and video, clearly two cash cow services, cannot be provisioned, and service-level agreements cannot be met.

To guarantee bandwidth, switches and routers must support priority queuing and offer some assurance that the queues will not be oversubscribed. Unfortunately, many vendors offer priority queuing, but because IP has no way to prevent oversubscription, they have no mechanism to complete the picture. A number of protocols have emerged that address this concern and will be discussed later in this chapter.

Two major QoS mechanisms have emerged for controlling quality guarantees across the network. The first is called *Explicit QoS*, in which applications have the capability to request specific levels of service quality. Routers or switches are then expected to make every effort possible to accommodate the request. The *Resource Reservation Protocol* (RSVP), discussed later, is a good example of an Explicit QoS technique.

The second approach is called *Implicit QoS*, in which routers and switches automatically allocate QoS resources such as bandwidth based on criteria established by a network administrator. These criteria include such things as the application type, logical port address, or protocol ID. The *Differential Services Working Group* (DiffServ) and ATM, also discussed later, are good examples of Implicit QoS.

To date, ATM has assumed the role as the ultimate QoS provider across the wide area. We have also noted, however, that it is not necessarily the ultimate long-term choice because of such factors as its tendency to be expensive, overhead-heavy, and somewhat complex. The ultimate question in all of this is whether QoS is sustainable across a *wide area network* (WAN) using protocols other than ATM. The answer, of course, is yes, although there can be a significant cost associated with doing so. The cost is often paid in bandwidth because to achieve sustainable QoS, networks have to be either overengineered or undersubscribed since protocols for intelligently provisioning and guaranteeing QoS are not quite ready for prime time. A great deal of discussion is going on these days in the trade journals by technicians who claim that the glut of bandwidth that is upon us makes it feasible to overengineer without a sense of waste. Although this is true from a brute force point of view, the philosophy is flawed. Parkinson's Law is alive and well in telecommunications, and although an overabundance of bandwidth is prevalent today, that will not be the case in the future. Thus, throwing bandwidth at the problem as a solution is less than ideal, and far from elegant. ATM provides the elegance today; other protocols will inherit the crown over time.

Quality is affected by a number of factors including latency or overall delay, echo cancellation, call-completion ratios, and post-dial delay, to name a few. One way around the challenge of dealing with the unpredictability of the Internet is to use a general technique called *Assured Quality Routing* (AQR). AQR, a term coined by ITSP iBasis, enables *voice over IP* (VoIP) carriers to transport voice traffic across the Internet until congestion results in unacceptable levels of quality, at which time AQR kicks in and routes calls off to the PSTN. This hybrid approach to voice is ideal for the service provider because it enables them to keep their costs down through the use of the Internet while enabling them to guarantee quality by using the PSTN as a fallback mechanism. The customer doesn't know this, nor do they need to know it. Once again, we see the magic of the opaque cloud.

In addition to those mentioned earlier, service providers must also consider the following factors when assessing the QoS level of the services they provide:

- *Scalability*: To what degree is their technology infrastructure scalable in terms of shelf space and ports?
- *Performance*: Does the infrastructure guarantee low packet delays and minimal jitters on an end-to-end basis?
- *Service quality*: Which standards does the manufacturer claim to abide by? Are they ISO 9000-certified?
- *Reliability*: How resilient is the hardware platform? Does it guarantee "five nines" of reliability (99.999 percent uptime)?
- *Management capability*: What OSS does the system support? Does it support such critical functions as billing and reconfiguration?
- *Interworking*: To what degree does the hardware/software package support the capability to effortlessly and seamlessly migrate between circuit and packet infrastructures?
- *Signaling*: What protocols are supported? Q.931, SS7, MF?

SERVICE AND PROTOCOL INTEROPERABILITY. The control protocols used to maintain QoS for VoIP and other packet-based services can be divided into three primary groups. The first of the three controls *network-to-network interworking*. In the PSTN, this is a central issue and has historically been one of the principal considerations of telephone company design engineers. Because of the large number of service providers and manufacturers active in the industry today, it is critical that some form of network-to-network interface be defined that enables an information exchange between providers and between equipment manufactured by distinct vendors. Although this capability is well defined in the PSTN, it is not yet well defined in the IP realm. Thus, for a provider to offer true IP telephony on a universal, full-service basis, the network-to-network control protocols must be clearly defined and distributed among all players.

The second of these is *device-to-device communications*, sometimes referred to as *peer-to-peer communications*. In most cases, the network assumes that in an IP environment, client devices will have some level of intelligence and the capability to establish sessions with other devices on an end-to-end basis either directly or by first establishing a session with some sort of mediation device such as a gateway or gatekeeper. Because these devices are intelligent, they can provide enhanced features such as CLASS, multiparty calling, bandwidth management, and so on. Included here is system-to-system messaging, which addresses billing, directory interoperability, provisioning, and inter-platform messaging. Finally, client-to-system messaging is important because it takes interoperability down to the customer level. Client-to-system messaging includes directory access, access to messaging applications, and client access to certain network features. A number of protocols are driving interoperability development, including the *Lightweight Directory Access Protocol* (LDAP), *Internet Message Access protocol* (IMAP), the *Extensible Markup Language* (XML), the *Wireless Access Protocol* (WAP), and a number of others.

The third area is *device management*, which encompasses both the control and content functions of VoIP service. The control function hosts the intelligence required to carry out telephony functions, while the content component carries out the mandates of the control component. For the purposes of IP telephony, a mechanism must perform call control, channel specifications, and CODEC configurations. To a large extent, these have been addressed by protocols such as H.323, SIP, and MGCP.

The protocols used for all three of these functions, from the *local area network* (LAN) to the WAN and back again, are described next.

QUALITY OF SERVICE (QoS) MANAGEMENT IN LOCAL AREA NETWORKS (LANS)

QoS begins at the customer location and is ideally maintained through the wide area to the receiving user device. It is important therefore that we begin our discussion of QoS with LANs.

Because of the migration to switched architectures in the LAN environment, the *Institute of Electrical and Electronics Engineers* (IEEE) released a standard in late 1998 for the creation of *virtual*

LANs (VLANs). A VLAN is nothing more than a single LAN that has been logically segmented into user groups or functional organizations. The elegance of the VLAN concept is that the members of a particular VLAN do not have to be physically collocated on the same network segment. They can be scattered all over the world as long as they are part of the same corporate network.

QoS is necessary in the LAN environment to overcome the disparity that often exists between the bandwidth offered by the LAN and the capacity of the WAN with which it communicates. LAN bandwidth has always been greater than that of WANs, and with the arrival of Fast and Gigabit Ethernet, the discrepancy between the two is widening at a rapid rate. Furthermore, the volume of LAN traffic is growing quickly, which means that the number of LANs contending for scarce WAN bandwidth is growing. Additionally, growth in IP-based networks, especially VPNs, is forcing that demand upward.

Several IEEE standards address the concerns of LAN QoS including *802.1Q* and *802.1p*. Both are discussed next.

802.1Q

The 802.1Q standard defines the interoperability requirements for vendors of LAN equipment wishing to offer VLAN capabilities. The standard was crafted to simplify the automation, configuration, and management of VLANs, regardless of the switch or end station vendor.

Like MPLS, 802.1Q relies on the use of priority tags that indicate service classes within the LAN. These tags form part of the frame header and use three bits to uniquely identify eight service classes. These classes, as proposed by the IEEE, are shown in Table 5-1.

802.1P

Closely associated with 802.1Q is 802.1p,[3] which enables the three QoS bits in the 802.1Q VLAN header to specify QoS

---

[3]The fact that the Q is upper case and the p lower case is not an accident. IEEE 802.1p is an adjunct standard and does not stand on its own. IEEE 802.1Q, on the other hand, is an independent standard.

TABLE 3-1   IEEE Service Classes

Priority	Binary Value	Traffic Type
7	111	Network Control
6	110	Interactive Voice
5	101	Interactive Multimedia
4	100	Streaming Multimedia
3	011	Excellent Effort
2	010	Spare
1	001	Background

requirements. It is primarily used by layer-2 bridges to filter and prioritize multicast traffic. The 802.1p QoS bits can be set by intelligence in the client machine, as dictated by the network policy established by the network management organization. In a practical application, 802.1p can be converted to DiffServ for QoS integration across the wide area. After all, 802.1p is really a QoS specification for LAN environments, typically Ethernet. Therefore, the DiffServ byte in the IP header can be encoded at the edge of the network by the ingress router, based on information contained in the 802.1p field in the Ethernet frame header. At the egress router, the opposite occurs, guaranteeing end to end QoS across the wide area.

Of course, these standards only address the requirements of LANs, which will inevitably interconnect with WAN protocols such as ATM or IP. Consequently, the IETF DiffServ committee has developed standards for interoperability between 802.1Q and wide area protocols such as IP's DiffServ, while the ATM Forum has a similar effort underway to map 802.1Q priority levels to ATM service classes.

The Differential Services Working Group (DiffServ) and the Multiprotocol Label Switching Working Group (MPLS)

The *Internet Engineering Task Force* (IETF) has taken an active role in the development of QoS standards for IP-based transmissions.

Under their auspices, two working groups have emerged with responsibilities for QoS issues. The first is the DiffServ and the second is MPLS.

When establishing connections for VoIP, it is critical to manage queues to ensure the proper treatment of packets that derive from delay-sensitive services. In order to do this, packets must be differentiable; that is, voice and video packets must be identifiable so that they can be treated properly. Routers in turn need to be able to respond properly to delay-sensitive traffic by implementing queue management processes. These require that the routers establish both normal and expedited queues and that they handle traffic in expedited routing queues faster than the arrival rate of the traffic. This translates into a traffic policing requirement to ensure that the offered load remains below the bandwidth reserved at each node for high-priority data.

DiffServ and MPLS have the same goal but approach it from different directions. DiffServ has the capability to prioritize packets through the use of bits in the IP header known as the *Differential Services Code Point* (DSCP), formerly part of the *Type of Service* (TOS) field. It relies on *Per Hob Behaviors* (PHBs) that define the traffic characteristics that must be accommodated. The best known of these is the *Expedited Forwarding PHB*, which is designed to be used for services that require minimum delays and jitters such as voice and video. DiffServ, then, is a technique for classifying packets according to QoS requirements. Because the classification process occurs at the edge of the network, it scales well as the network grows.

DiffServ breaks the responsibilities of traffic management into four key areas based on the overall architecture of the network, as illustrated in Figure 5-4. At the customer's access router, traffic is managed according to flow requirements and clustered for delivery to the service provider's network. The traffic is then handed to the service provider's ingress router, which sits at the edge of the network and is responsible for implementing the service-level agreement between the customer and the service provider.

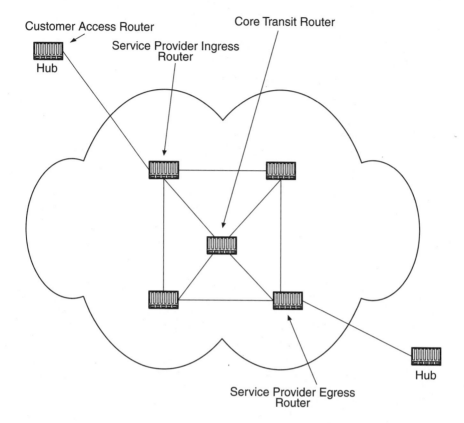

FIGURE 3-4   A DiffServ network

Once the edge routers have classified the traffic, the core routers can handle it according to the DSCP "markers" they assign to each packet. Within the network, then, core transit routers simply route the traffic as required. By the time the traffic arrives at the core, the edge devices have already classified it, and the core router simply handles the interior routing function. Ultimately, the traffic reaches the service provider's egress router, another edge device, which performs additional traffic-shaping functions to ensure compliance with the service-level agreement. Thus, the core can be extremely fast because the classification process has already been done at the point of ingress.

DiffServ, therefore, is *not* an end-to-end protocol. Traffic, perhaps from a LAN, arrives at the edge of the network, where DiffServ's domain begins. The ingress and egress routers manage and shape traffic flows with the freedom to discard packets if appropriate.

MPLS is considered to be superior to DiffServ, although both techniques rely on edge routers to classify and label the packets at the point of ingress. MPLS achieves the same goal as DiffServ by establishing virtual circuits known as *Label Switched Paths* (LSPs) which are built around specific QoS requirements. Thus a router can establish LSPs with explicit QoS capabilities and route packets to those LSPs as required, guaranteeing the delay that a particular flow will encounter end-to-end. Some industry analysts have compared MPLS LSPs to the trunks that are established in the voice environment.

While both MPLS and DiffServ offer reasonable levels of QoS control in IP environments, neither has the level of capability that ATM offers, which explains why ATM combined with IP is the choice of technologies for many players looking to deploy mixed services over IP. Both techniques offer services to the user, but they are lacking in a number of key QoS areas such as routing capability. Both DiffServ and MPLS describe techniques for classifying and labeling a variety of QoS levels, but neither of them speak to the requirements for establishing an end-to-end path that offer constant quality. So while both (and especially MPLS for its simplicity) are often compared to ATM, ATM is clearly a more robust and capable protocol. *Multiprotocol Over ATM* (MPOA) is one technique that ATM uses to manage ingress traffic and identify flows that require diverse QoS levels. It establishes virtual channels for specific QoS between various devices within the network, and routes traffic as befits its unique requirements for service. If it finds traffic that requires specific handling that does not already have a virtual channel established, it creates a default channel to ensure delivery. Thus, ATM handles both packet classification *and* routing.

MPLS is similar to MPOA and uses a two-part process for routing. First, it divides the packets into various *Forwarding Equivalence Classes* (FECs) and then maps the FECs to their next hop point. This process is performed at the point of ingress. Each FEC is given a fixed-length "label" that accompanies each packet from hop to hop. At each router, the FEC label is examined and used to route the packet to the next hop point where it is assigned a new label.

INTEGRATED SERVICES (INTSERV) AND THE RESOURCE RESERVATION PROTOCOL (RSVP)

Although *Integrated Services* (IntServ) and the *Resource Reservation Protocol* (RSVP) seem to be waning in popularity as broadly accepted QoS techniques, they are worth discussing.

IntServ is primarily seen as a useful QoS mechanism for the edge of the network and is divided into two mechanisms. The first mechanism consists of devices that have the capability to differentiate between various QoS levels so that data with differing requirements can be properly handled per the stipulations of the service-level agreement. The second piece is a notification mechanism that enables an application-requesting service to set up a path that will in fact guarantee that service on an end-to-end basis.

Let's now examine how IntServ works. When a host requests a specific QoS level for its data stream, RSVP transports the setup request through the network on a node-by-node basis. At each router, the RSVP setup packet attempts to reserve resources if they are available in order to maintain the QoS state until the transmission is complete. What arrives at the receiver is a rather large packet that contains path information that describes the QoS that can be expected from that path. The receiving application examines the information contained in the RSVP setup packet, requests performance parameters, and hands the information back to RSVP, which transmits the information upstream to the originating device. The device hands the packet to the sending application, notifying it of the level of service it can expect across the requested path. The routers along the way make every effort to satisfy the request for service.

RSVP, then, is a flow-based protocol that enables an application to communicate its QoS requirements on an end-to-end basis, stipulating the bandwidth, latency, and jitter requirements that must be met for the application at hand.

A primary assumption made by RSVP is that it will largely be used for video and other high-speed multicast applications. It therefore attempts to inject a modicum of determinism in a connectionless network so that multimedia streams can be transported with acceptable levels of transmission quality.

RSVP's greatest shortcoming is its lack of scalability. It performs well when implemented in small networks but falls short when used in high-volume, wide area situations. In order for it to work properly, RSVP requires the establishment of end-to-end paths with guaranteed quality. In a small, private network, this can be done, but when the Internet is the WAN, it is simply impossible to maintain a guaranteed end-to-end quality state today. RSVP also places a significant processing load on routers, which can cause the service quality to be downgraded.

In summary, RSVP is not a routing protocol. That function is performed by other protocols such as IP, *Open Shortest Path First* (OSPF), *Routing Information Protocol* (RIP), *Border gateway Protocol* (BGP4), and Enhanced IGRP. Its sole responsibility is to ensure that resources are available for a particular QoS level before enabling a session to be established that requires that level of quality.

SIGNALING IN VOIP ENVIRONMENTS

Telephony, whether circuit-switched or IP-based, requires both basic and supplementary (sometimes called enhanced) services. The basic services include call management, call authentication, caller identification, security, billing, network management, and bandwidth assignment capabilities. Supplementary services include unified messaging, integrated voice response, 911, 800 number service, conference calling, online directory services, and many of the CLASS services that SS7 makes possible. In this section, we will discuss the most common signaling protocols that have been proposed for the IP telephony domain. It is important to recognize that in the PSTN, SS7 enables the signaling information and the actual content to be transported across separate network infrastructures. In IP, however, the two share the same network. In the IP domain, protocols such as H.323 or SIP are responsible for signaling, including call setup and device mapping. MGCP and SS7 control the interface to gateways and call agents. The content, then, is governed by protocols such as the *Real-Time Transport Protocol* (RTP).

H.323

H.323 started as H.320 in 1996, an *International Telecommunications Union* (ITU) standard for the transmission of multimedia content over ISDN. Its original goal was to connect LAN-based multimedia systems to network-based ones. It defined a network architecture that included gatekeepers that performed zone management and address conversions, endpoints that were terminals and gateway devices, and multimedia control units that served as bridges between multimedia types.

H.323 has been rolled out in four phases. Phase one defines a three-stage call setup process. The first stage is comprised of a pre-call step where user registration, connection admission, and an exchange of status messages that are required for the call setup are performed. Next is the actual call setup process, which uses messages similar to ISDN's Q.931. Finally, the capability exchange stage establishes a logical communications channel between the communicating devices and the identified conference management details.

Phase two enables the use of the *Real-Time Transport Protocol* (RTP) over ATM. This eliminates the added redundancy of IP and also provides privacy, authentication, and greatly demanded telephony features such as call transfers and call forwarding. RTP also has an added advantage. When errors result in a packet loss, RTP does not request resends of those packets, thus providing for real-time processing of application-related content. No delays result from errors.

Phase three adds the capability of transmitting real-time faxes after establishing a voice connection. Phase four, released in May 1999, adds call connections over the *User Datagram Protocol* (UDP). This significantly reduces call setup times; inter-zone communications; call holds, parks, and pickups; and call and message waiting features. This last phase bridges the considerable gap between IP voice and IP telephony.

One concern with H.323 is interoperability. Currently, three versions are available, and because of a lack of deployment coordination, all three have a significant installed base that naturally leads to a certain amount of fragmentation. Consequently, interoperability has become a serious concern. One solution that has been offered is the iNOW! Profile (www.imtc. org/act_inow.htm). Developed initially by Lucent, VocalTec, and ITXC, iNOW! was created to define the options that manufacturers can use to ensure interoperability for multimedia, voice, and fax. INOW! recently merged with the *International Multimedia Teleconferencing Consortium* (IMTC) whose stated goals are to sponsor and conduct interoperability tests between suppliers of conferencing products and services, provide a forum for technical exchanges between IMTC members that will lead them to make additional submissions to the standards bodies to support interoperability and usability of multimedia teleconferencing products and services, and finally to educate the business and consumer communities on the benefits of the underlying technologies and applications. Under the auspices of IMTC, iNOW! has the following responsibilities:

- To help members build interoperable IP telephony products based on the current and future IMTC iNOW! interoperability Profiles developed and approved by the Activity Group.
- To provide a forum for technical exchanges and to help with the resolution of interoperability technical issues that affect interoperability between iNOW! compliant products and other products.
- To define test strategies that member companies can use to test their products for interoperability.

- To support IMTC and *European Telecommunications Standards Institute* (ETSI) TIPHON testing of cross-vendor product testing.

- To support the development of additional features and a broader scope for IMTC iNOW! Profiles as the need arises.

Several Internet telephony interoperability concerns are addressed by H.323. These include gateway-to-gateway interoperability, which ensures that telephony can be accomplished between different vendors' gateways; gatekeeper-to-gatekeeper interoperability, which does the same thing for different vendors' gatekeeper devices; and finally gateway-to-gatekeeper interoperability, which completes the interoperability picture.

SESSION INITIALIZATION PROTOCOL (SIP)

Although H.323 has its share of supporters, it is slowly being edged out of the limelight by the IETF's SIP. SIP supporters claim that H.323 is far too complex and rigid to serve as a standard for basic telephony setup requirements, arguing that SIP, which is architecturally simpler and imminently extensible, is a better choice. In reality, H.323 is an umbrella standard that includes (among others) H.225 for call handling, H.245 for call control, G.711 and G.721 for CODEC definitions, and T.120 for data conferencing. Originally created as a technique for transporting multimedia traffic over a LAN, gatekeeper functions have been added that enable LAN traffic and LAN capacity to be monitored so that calls are established only if adequate capacity is available on the network. Later, the *Gatekeeper Routed Model* was added, enabling the gatekeeper to play an active role in the actual call setup process. This meant that H.323 had migrated from being purely a peer-to-peer protocol to having a more traditional, hierarchical design.

The greatest advantage that H.323 offers is maturity. It has been available for some time now and, although robust and full-featured, was not originally designed to serve as a peer-to-peer protocol. Its maturity therefore is not enough to carry it. It currently lacks a network-to-network interface, nor does it adequately support congestion control. This is not generally a problem for private networks, but it does become problematic for service providers who want to interconnect PSTNs and provide national service among a cluster of providers. As a result of this, many service providers have chosen to deploy SIP instead of H.323 in their national networks.

SIP is designed to establish peer-to-peer sessions between Internet routers. The protocol defines a variety of server types, including feature servers, registration servers, and redirect servers. SIP supports fully distributed services that reside in the actual user devices, and because it is based on existing IETF protocols, it provides a seamless integration path for voice/data integration.

Similar to H.323, SIP does not yet offer a network-to-network interface, but the IETF has created a working group designed to bring together the best features of SIP and *MeGaCo* (an important protocol that will be discussed in detail in the next section) to overcome this obstacle.

Ultimately, telecommunications, like any industry, revolves around profitability. Any protocol that enables new services to be deployed inexpensively and quickly immediately catches the eye of service providers. Like TCP/IP, SIP provides an open architecture that can be used by any vendor to develop products, thus ensuring multivendor interoperability. And because SIP has been adopted by such powerhouses as Lucent, Nortel, Cisco, Ericsson, and 3Com and is designed for use in large carrier networks with potentially millions of ports, its success is reasonably assured.

Originally, H.323 was to be the protocol of choice to make this possible. Although H.323 is clearly a capable suite of protocols and is indeed quite good for VoIP services that derive from ISDN implementations, it is still incomplete and is quite complex. As a result, it has been relegated to use as a video control protocol and for some gatekeeper-to-gatekeeper communications functions.

The intense interest in moving voice to an IP infrastructure is driven by simple and understandable factors: cost of service and enhanced flexibility. However, in keeping with the "Jurassic Park Effect" (Just because you *can*, doesn't necessarily mean you *should*.), it is critical to understand the differences that exist between simple voice and full-blown telephony with its many enhanced features. It is the feature set that gives voice its range of capabilities. A typical local switch such as Lucent Technologies' 5ESS offers more than 3,000 features, and more will certainly follow. Of course, the features and services are possible because of the protocols that have been developed to provide them across an IP infrastructure.

MEDIA GATEWAY CONTROL PROTOCOL (MGCP) AND FRIENDS

Many of the protocols that are guiding the successful development of VoIP efforts today stem from work performed early on by Level 3 and Telcordia, which together founded an organization called the International SoftSwitch Consortium. In 1998, Level 3 brought together a collection of vendors who collaboratively developed and released the *Internet Protocol Device Control* (IPDC). At the same time, Telcordia created and released the *Simple Gateway Control Protocol* (SGCP). The two later merged to form the *Media Gateway Control Protocol* (MGCP), discussed in detail in RFC 2705.

MGCP enables a network device responsible for establishing calls to control the devices that actually perform IP voice streaming. It permits software call agents and media gateway controllers to control streaming media gateways at the edge of the network. These gateways can be cable modems, set-top boxes, *private branch exchanges* (PBXs), *Voice and Telephony over ATM* (VTOA) gateways, and VoIP gateways. Under this design, the gateways manage the circuit-switch-to-IP voice conversion, while the agents manage signaling and call processing.

MGCP makes the assumption that call control in the network is software-based and resides in external intelligent devices that perform all call control functions. It also makes the assumption that these devices will communicate with one another in a primary-secondary arrangement under which the call agents send instructions to the gateways for execution. Table 5-2 displays the bandwidth requirements and delay sensitivity for a variety of different applications.

Meanwhile, Lucent created a new protocol called the *Media Device Control Protocol* (MDCP). The best features of the original three were combined to create a full-featured protocol called the MeGaCo, also defined as H.248. In March of 1999, the IETF and ITU met collaboratively and created a formal technical agreement between the two organizations, which resulted in a single protocol with two names. The IETF calls it MeGaCo; the ITU calls it H.GCP.

MeGaCo/H.GCP operates under the assumption that network intelligence is housed in the central office and therefore replaces the gatekeeper concept proposed by H.323. By managing multiple gateways within a single IP-equipped central office, MeGaCo minimizes the complexity of the telephone network. In other words, a corporation might be connected to an IP-capable *central office* (CO), but because of the IP-capable switches in the CO that have the capability to convert between circuit-switched and packet-switched voice, full telephony features are possible. Thus, the next-generation switch converts between circuit and packet, while MeGaCo performs the signaling necessary to establish a call across an IP WAN. It effectively bridges the gap between legacy SS7 signaling and the new requirements of IP while supporting both connection-oriented and connectionless services.

TABLE 5-2 Bandwith Requirements and Delay Sensitivity for Common Applications

APPLICATION	REQUIRED BANDWIDTH	SENSITIVITY TO DELAY
Voice	Low	High
Video	High	Medium to high
Medical imaging	Medium to high	Low
Web surfing	Medium	Medium
LAN interconnection	Low to high	Low
E-mail	Low	Low

IP TELEPHONY SUMMARY

Let's summarize, then, how a telephone call is carried across an IP network. A customer begins the call using a traditional telephone. The call is carried across the PSTN to an IP telephony gateway, which is nothing more than a special purpose router designed to interface between the PSTN and a packet-based IP network. As soon as the gateway receives the call, it interrogates an associated gatekeeper device that provides information about billing, authorization, authentication, and call routing. As soon as the gatekeeper has delivered the information to the gateway, the gateway transmits the call across the IP network to another gateway, which in turn hands the call off to the local PSTN at the receiving end, completing the call.

IP telephony is primarily served by two protocols, H.323 and MGCP. H.323, from the ITU, is a peer-to-peer protocol under which the station initiating the call and the station terminating the call are peers with one another. The protocol often defines a gatekeeper responsible for address translation and zone management, but at the same time it requires that gateway devices and terminals be responsible for call control and call processing functions. As a call control protocol, H.323 is most capable when implemented in an environment in which the end stations are intelligent, such as PC telephony client devices and integrated PBX/gateways. Because of its complexity, however, H.323 is being pressured by the far simpler SIP protocol.

MGCP, on the other hand, is an IETF protocol born of the melding of Telcordia's SGCP and IPDC. Originally designed for use in gateway devices, it is now often found in client devices, and although designed for telephony, it also serves well as a multimedia control protocol.

Voice Over IP (VoIP) Implementation Summary

There is no question that IP voice has become a reasonable technological alternative to traditional PCM voice. VoIP gateways are in various stages of development with regard to reliability, features, and manageability. Consequently, service providers interested in deploying VoIP solutions have a number of possible options available to them. One is to accept the current state of the technology and deploy it today in a trial mode while waiting for enhanced versions to arrive. This would gain service providers an early lead but would probably be a problem in that it would lock them into an inferior technology that would ultimately be a setback.

A second option is a variation of the first: implement the technology that is available today, but make no guarantees on service quality. This approach is currently being used in Western Europe and in the southwestern U.S. as a way to provide low-cost, long-distance service to communities there. The service is actually quite good, and although it is not "toll quality," it is inexpensive. The marketers of the service make it clear at the time of the sale that the quality will often not be equivalent to that of the traditional telephone network so there is no danger of misrepresentation. Furthermore, the companies deploying the service have no billing infrastructure to support because they rely exclusively on prepaid calling cards; thus, they can keep their costs low.

Although many believe that IP telephony will fail due to its perceived lower service quality, others point to cellular telephony, the success of which has far exceeded all expectations in spite of the inferior quality that it delivers. Clearly, then, the market will accept lower than toll quality voice if there is perceived value in the trade-off.

VoIP faces a number of challenges in the marketplace today. These include the proprietary nature of gatekeeper software; the fact that many devices only offer partial compliance with H.323; a proliferation of competing standards bodies, including the IETF, the ITU, ETSI's TIPHON, and the IMTC; and actual implementations that demonstrate VoIP's capabilities with customers under realistic load conditions. Another challenge is the fact that many VoIP solutions today require the user to dial long strings of digits before their call can be connected. This does not represent a simplification of telephony due to the implementation of IP.

And the bottom line? Ultimately, IP's success, whether for voice or data, hinges on one clear fact: it will only be widely accepted if it has the capability to offer customers a true VPN that delivers all services requested over a single network interface with an expected high level of quality.

When implementing VoIP technology as a migration path, corporations should be sensitive to such obvious concerns as QoS, billing, network management, and security, but they should also question the efficacy of such issues as short- and long-term costs, scalability, standards compliance, and interoperability.

NETWORK MANAGEMENT FOR QoS

Because of the diverse audiences that require network performance information and the importance of *service-level agreements* (SLAs), the data collected by network management systems must be malleable so that it can be formatted for different sets of corporate eyes. For the purposes of monitoring performance relative to SLAs, customers require information that details the performance of the network relative to the requirements of their applications. For network operations personnel, reports must be generated that detail network performance to ensure that the network is meeting the requirements of the SLAs that exist between the service provider and the customer. Finally, reports must be available to enable sales and marketing organizations to properly represent the company's offerings to customers and enable them to anticipate requirements for network augmentation and growth.

For the longest time, the *Telecommunications Management Network* (TMN) has been considered the ideal model for network management. As network profiles have changed, however, with the steady migration to IP and a renewed focus on service rather than technology, the standard TMN philosophy has begun to appear somewhat tarnished.

Originally designed by the ITU, TMN is built around the *Open Systems Interconnection* (OSI) model and its attendant standards, which include the *Common Management Information Protocol* (CMIP) and the *Guidelines for the Development of*

Managed Objects (GDMO). TMN employs a model, shown in Figure 5-5, comprising a *network element layer,* an *element management layer,* a *network management layer,* a *service management layer,* and a *business management layer.* Each has a specific set of responsibilities closely related to those of the layers that surround it.

The network element layer defines each manageable element in the network on a device-by-device basis. Thus, the manageable characteristics of each device in the network are defined at this functional layer.

The element management layer manages the characteristics of the elements defined by the network element layer. The information found here includes activity log data for each ele-

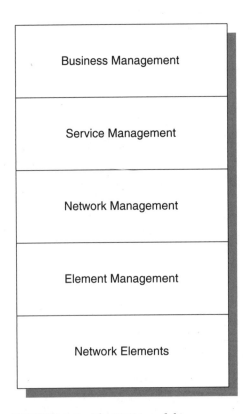

FIGURE 5-5 The TMN model

ment. This layer houses the actual element management systems responsible for the management of each device or set of devices in the network.

The network management layer has the capability to monitor the entire network based upon information provided by the element management layer.

The services management layer responds to information provided by the network management layer to deliver such service functions as accounting, provisioning, fault management, configuration, and security services.

Finally, the business management layer manages the applications and processes that provide strategic business planning and the tracking of customer interaction vehicles such as SLAs.

OSI, while highly capable, has long been considered less efficient than IETF management standards and in 1991 market forces began to shift. That year, the Object Management Group was founded by a number of computer companies including Sun, Hewlett-Packard, and 3Com. Together they introduced the *Common Object Request Broker Architecture* (CORBA). CORBA is designed to be vendor-independent and built around object-oriented applications. It enables disparate applications to communicate with each other, regardless of physical location or vendor. Although CORBA did not achieve immediate success, it is now widely accepted, resulting in CORBA-based development efforts among network management system vendors. This may seem to fly in the face of the OSI-centric TMN architecture, but it really doesn't. TMN is more of a philosophical approach to network management and does not specify technological implementation requirements. Thus, a conversion from CMIP to the *Simple Network Management Protocol* (SNMP) or the implementation of CORBA does not affect the overall goal of the TMN.

IMPLEMENTING IP SERVICES

So what does this all mean? Consider the directions that major manufacturers are taking today with regard to IP migration.

Lucent Technologies and Ericsson, both critical players in the IP telephony marketplace, have targeted carriers with new gateway products that use *Automatic Number Identification* (ANI) to select IP voice and eliminate the secondary dial tone that characterized early VoIP systems. Nortel Networks offers single-stage dialing (no secondary dial tone) and embedded CLASS services. Some vendors have offered customers the ability to select IP voice by simply dialing an access code: "Dial nine for a standard outside line, seven for a voice-over-IP line." Lucent Technologies, working with VocalTec, has developed interoperable gateway devices fully compatible with the mandates of the iNOW! Profile. Motorola, working closely with NetSpeak, offers wireless IP voice services, and others will certainly follow.

It would appear that IP convergence will manifest itself in six significant ways: in carrier-class IP voice, in the growth of *Internet telephony service providers* (ITSPs), in the expanded use of private corporate gateways, in voice-enabled Web sites, in IP-enabled call centers, and in *Application Service Providers* (ASPs).

CARRIER CLASS IP VOICE

According to the International Data Corporation, domestic revenues for VoIP will grow at a compound annual rate of 103.4 percent between 1997 and 2002 to $24.39 billion. Internationally, growth will reach 100.9 percent in the same period, cresting at $20.49 billion in revenues.

Corporations that are considering a conversion to VoIP should take into account the following concerns when developing their migration strategies:

- Because standards compliance remains fragmented in the VoIP domain, corporations must be careful when selecting hardware and software platforms to ensure that they do not lock themselves into a proprietary solution. Similarly, implementers must ensure that vendor-to-vendor interoperability is assured because most IT organizations are loath to implement a single vendor solution.

- The requirements placed upon voice networks and data networks are unique. As they converge to a common infrastructure, steps should be taken to ensure that the requirements of both can be satisfied by the new fabric. Related to this is the requirement for capable management solutions. To guarantee the QoS that negotiated SLAs demand, network managers must have access to a capable network management system that reflects the information they require to ensure SLA compliance.

From a customer's perspective, the principle advantages of VoIP include consolidated voice, data, and multimedia transports; the elimination of parallel systems; the ability to exercise PSTN fallback in the event that the IP network becomes overly congested; and the reduction of long-distance voice and facsimile costs.

For an *Internet service provider* (ISP) or a *Competitive Local Exchange Carrier* (CLEC), the advantages are different but no less dramatic, including the efficient use of network investments due to traffic consolidation, new revenue sources from existing clients because of demands for service-oriented applications that benefit from being offered over an IP infrastructure, and the option of transaction-based billing. These can collectively be reduced to the general category of customer service, which service providers such as ISPs and CLECs should be focused on. The challenge they face will be to prove that the service quality they provide over their IP networks will be identical to that provided over traditional circuit-switched technology.

Major carriers are voting on IP with their own wallets, a sure sign of confidence in the technology. In June of 1999, Level 3 became the first carrier to offer voice and data services across an all-IP network using Lucent's software-based SoftSwitch switching platform. Shortly thereafter, Frontier Corporation announced an agreement with Lucent Technologies worth several hundred million dollars to use the same technology in the core of their all-

IP network. By the end of 2002, the company plans to shift $3 billion in service revenues to the IP backbone and expects to see capital equipment costs go down by as much as 50 percent, with equipment space and power costs reduced by as much as 75 percent. The SoftSwitch platform that both companies will use also offers integrated operation and application support for billing, AIN-like service creation, and customer support.

These IP-based software switches enable call-processing and switching functions to be distributed across a series of integrated servers that enable an extremely fine control of availability and scalability. Lucent Technologies and Nortel Networks have entered the fray with their own versions of "decomposed switch platforms," offering their 7 R/E and Succession Network products, respectively. Both comprise a collection of high-end servers interconnected via IP or ATM that facilitate the evolution from today's narrowband network to a broadband platform that offers distributed access and control to deliver a wide variety of integrated services. In effect, these devices migrate central office functions from the core to the edge of the network. This distribution process dramatically lowers the cost of entry and adds enormous flexibility to the service provider's abilities by providing a bridge between the circuit- and packet-switched domains. Ultimately, these devices are designed to not only distribute the switching function, but also to reduce the number of network elements required to provide switched services, to ensure interoperability and service flexibility, to guarantee scalability, and to deliver carrier-class voice with "five nines" of reliability.[4]

As we observed earlier, the key to IP's success in the voice provisioning arena lies with its "invisibility." If done correctly, service providers can add IP to their networks, maintaining service quality while dramatically improving their overall efficiency. IP voice (not to be confused with Internet voice) will be implemented by carriers and corporations as a way to reduce

[4]Meaning that the devices will be available 99.999 percent of the time. That works out to less than six minutes of downtime per year.

costs and move to a multiservice network fabric. Lucent, Nortel, and Cisco have all added SS7 and IP voice capabilities to their router products and access devices, recognizing that their primary customers are looking for IP solutions. *Digital Switch Corporation* (DSC) has demonstrated AIN applications in an IP setting, and as a demonstration of its commitment to voice as a strategic payload, Cisco purchased Summa Four, the manufacturer of an IP voice switch capable of handling as many as 1,000 simultaneous IP voice calls. Qwest and Level 3 are fully committed to IP, and equipment manufacturers are providing the hardware necessary to meet their requirements. Thus, voice delivered across an IP infrastructure is a desirable component of the modern service provider's arsenal.

INTERNET TELEPHONY SERVICE PROVIDERS (ITSPs) AND PRIVATE CORPORATE GATEWAYS

Internet telephony offers revenue opportunities for backbone ISPs and legacy circuit-switched service providers that want to either enter the voice business or reduce their costs of providing service. Because of the inherently efficient nature of packet-based transmission, IP telephony represents a real opportunity for service providers to dramatically increase the capacity of their voice networks.

IP telephony, however, has a downside that cannot be ignored. A significant difference exists between running an ISP and running a telephone company, just as a difference exists between VoIP and IP telephony. Although Internet telephony is IP telephony, IP telephony is *not* necessarily Internet telephony. Many corporations have built internal IP networks that carry multimedia traffic including voice. Because the networks are privately owned and operated, the IT staff responsible for service quality can regulate traffic across the backbone, employ proprietary technologies and products to ensure QoS, fall back on PSTN if appropriate, and, to a large extent, manage user expectations. Thus, IP can provide "toll quality" service if deployed properly. That model, however, is a far cry from Internet telephony. Today the Internet can certainly deliver IP voice, but it cannot be relied upon to deliver sustainable QoS. Thus, any attempt in the public domain to replace circuit-switched voice with Internet telephony is doomed to face serious marketplace resistance because of the (correct) perception of inferior QoS. Until QoS protocols such as those discussed earlier become widespread and dependably functional, Internet telephony will be forced to coexist with circuit switching, which will remain viable for some time to come. Under certain circumstances, however, corporate IP gateways for voice are quite viable and are widely used today.

CLEARINGHOUSE SERVICES

An emerging aspect of IP telephony is the concept of *clearinghouse services*. ITSPs can form relationships with clearinghouses, which save them the effort and expense associated with forming individual relationships with myriad other ITSPs they must interface with in order to be able to hand off calls in areas where they want to provide service. Instead, the clearinghouse performs this function for the ITSP. The ITSP establishes a single interface with the clearinghouse, which then manages all call routing, authorization, billing, and network management for the ITSP. ITXC provides a clearinghouse service, as do AT&T's Global Clearinghouse and Arbinet.

Some ITSPs also employ post-paid calling cards. Instead of the customer purchasing a pre-paid card, the customer provides the service provider with a credit card number that calls are billed to once the customer provides the appropriate PIN. Thus, the billing is done after the call. Clearinghouse services manage this process as well.

IP-ENABLED CALL CENTERS

Ultimately, call centers are nothing more than enormous routers. They receive incoming data delivered using a variety of media (phone, e-mail, facsimile transmissions, mail order) and make decisions about handling them based on information contained in each message. One challenge that has always faced call center management is the ability to integrate message types and route them to a particular agent based on specified decision-making criteria such as name, address, phone number, e-mail address, ANI triggers, product purchase history, the customer's geographic location, or language preference. This has resulted in the development of a technical philosophy called *unified messaging*. With unified messaging, all incoming messages for a particular agent, regardless of the media over which they are delivered, are housed centrally and clustered under a single account identifier. When the agent logs into the network, their PC lists all of the messages that have been received for them, giving them the ability to much more effectively manage the information contained in those messages.

Today unified messaging systems also support the road warriors. A traveling employee can dial into a message gateway and download all messages—voice, fax, or e-mail—from that one central location, thus dramatically simplifying the process of staying connected while away from the office.

Call centers are undergoing tremendous changes as the IP juggernaut hits them. The first of these is a redefinition of the market they serve. For the longest time, call centers have primarily served large corporations because they are expensive to deploy. With the arrival of IP, however, the cost is dropping, and some industry authorities believe that by the end of 2000 the cost of a position in an IP call center will be approximately half that of a traditional call center.

The second major change is that call centers are becoming the focal point for corporations with effective and successful e-commerce applications. Small and medium-size businesses are enjoying an expanded market presence thanks to e-commerce, and the Internet and the disintermediation that it brings about are helping them lower their operational costs.

One customer complaint that is often voiced about the Web, particularly with regard to e-commerce, has to do with its lack of interactivity. When customers use the Web to make purchases, they are often faced with the dilemma of needing more information than is available at the Web site of the company they want to do business with, which often requires that they disconnect from their ISP and call the company's 800 number, a step that effectively obviates the need for ordering online. In response to this, some e-commerce providers have implemented the ability to invoke an Internet telephony session with an agent while product surfing. Although this feature is not yet as good as it soon will be, the fact that major corporations are adding voice capabilities to their Web sites and that customers are using these services proves that the demand for integrated voice and data must be satisfied.

INTEGRATING THE PBX

An enormous installed base of legacy PBX equipment exists, and vendors did not enter the IP game enthusiastically. Early entrants arrived with enhancements to existing equipment that were proprietary, expensive, and did little to raise customer awareness or engender trust in the vendor's migration strategy. Over time, however, PBX manufacturers began to embrace the concept of convergence as their customers' demands for IP-based systems grew, and soon products began to appear. Most have heard the message delivered by the customer: preserve the embedded base to the greatest degree possible as a way of saving the existing investment, create a migration strategy that is seamless and as transparent as possible, and preserve the features and functions that customers are already familiar with to minimize churn.

Major vendors like Lucent Technologies and Nortel Networks have responded with products that do exactly that. Lucent's DEFINITY IP Solutions package is an upgrade to their DEFINITY Enterprise Communications Server PBX that incorporates IP telephony capabilities. It enables voice calls and faxes to be carried over LANs, WANs, the Internet, and corporate intranets. It also functions as both a gateway and gatekeeper, providing circuit-to-packet conversions, security, and access to a wide variety of applications including enhanced call features such as multiple line appearances, hunt groups, multi-party conferencing, call forwarding holds, call transfers, and speed dial. It also provides access to voice mail, CTI applications, wireless interfaces, and call center features. An additional product, the DEFINITY IP SoftPhone, enables a PC to serve as a full-feature business telephone.

Nortel's Meridian Integrated IP Telephony Gateway is a card-based adjunct that is installed in the Meridian 1 Intelligent Peripheral Equipment shelf. It interconnects multiple systems across a private network, creating a VPN-like environment that can carry compressed voice and fax messages as IP packets across an IP network. The product is H.323-compliant and enables corporations to take advantage of the Meridian's well-known feature set including least cost routing, QoS discrimination, systems management, and discrete billing. It also supports the capability to fall back to the PSTN if QoS suffers due to congestion or packet loss.

Like Lucent's DEFINITY, the Meridian system supports the capability to carry IP telephony down to the desktop level. The Meridian IP Telecommuter software enhancement enables a PC to serve as a full-feature telephone set.

BILLING AS A CRITICAL SERVICE

One area that is often overlooked when companies look to improve the quality of the services they provide to their customers is billing. Although it is not typically viewed as a strategic competitive advantage, studies have shown that customers view it as one of the top considerations when assessing service provider abilities.

Billing offers the potential to strengthen customer relationships, improve long-term business health, cement customer loyalty, and generally make businesses more competitive. However, for billing to achieve its maximum benefit and strategic value, it must be fully integrated with a company's other support systems including network and service provisioning systems, installation support, repair, network management, sales, and marketing. If done properly, the billing system becomes an integral component of a service suite that enables the service provider to quickly and efficiently introduce new and improved services in logical bundles; improve business indicators such as service timeliness, billing accuracy, and cost; offer custom service programs to individual customers based on individual service profiles; and transparently migrate from legacy service platforms to the next-generation network.

In order for billing as a strategic service to work successfully, service providers must build a business plan and migration strategy that takes into account the integration with existing operations support systems, business process interaction, the role of IT personnel and processes, and post-implementation testing to ensure compliance with strategic goals stipulated at the beginning of the project.

APPLICATION SERVICE PROVIDERS (ASPs)

It is amazing how circular the telecommunications industry is. A few years ago, the industry announced a new technology called the *Digital Subscriber Line* (DSL), followed by another known as the VPN. DSL, of course, is a reworking of a legacy technology called *Data over Voice* (DOV), while VPN is an IP version of the old X.25 concept of *Close User Groups* (CUG). ASPs represent another reinvention, hearkening back to the days of timeshare systems and service bureaus.

An ASP is an application hosting service that enables companies to "rent" access to applications that they require to run their business operations. They are hosted at a central data center and are typically accessed via the Web or an application-specific client. Most ASPs offer such services as database backups, disaster recovery, application access, database access, and service-level agreement administration. The actual applications that are available from ASPs range from such mainstays as Microsoft Office to complex *enterprise resource planning* (ERP) applications.

ASPs provide numerous advantages, including improved customer service, reduced operations costs, and reduced risks due to the physical dissemination of computer and application resources. Although the advantages of using an ASP may seem obvious, corporations do not gain benefit from their use equally. In many cases, large corporations have already outsourced IT operations and applications, thus largely eliminating the need for an ASP's services. Small businesses require little in the way of IT resources beyond a small client-server LAN and in many cases can't afford the services of an ASP.

The mid-level market, however, is ideally positioned to benefit from the services offered by an ASP. Although they are larger than the small business tier of companies, they are large enough to need access to the same applications used by large corporations. They typically can't afford to own them, however, so a shared application environment offered by an ASP represents an ideal compromise for the medium-size enterprise.

Quite a few companies have announced their services as ASPs, including Interliant, InSynQ, Telecomputing, Corio, EDS, IBM Global Services, Groupe Bull, Agillion, Brightstar Information Technology Group, Interpath Communications, and Breakaway Solutions. Sprint has also entered the ASP business with flair and gusto, opening the first of 12 data centers that will serve as hosting sites for e-business solutions. In partnership with Deloitte and Touche, Sprint will offer a variety of online applications.

Similarly, Frontier announced its intention to enter the ASP business, offering to host e-mail and message outsourcing services for corporations. Oracle Corporation made minor history when it became the first software provider to create its own ASP and Qwest appears to be the most aggressive of all. After forging ASP alliances with more than 50 companies, Qwest actually competes with itself in some cases. Microsoft has announced that it will host its Office applications across the Web, and other players will certainly follow. International Data Corporation estimates that while the ASP industry is still small ($150 million in 1999), it will grow rapidly, reaching $2 billion by 2003. Forrester Research expects even greater performance, with ASP revenues reaching $6 billion by 2001.

Of course, the ASP model has a downside, which is the degree to which corporations are dependent upon it for the delivery of business-critical application and database hosting services. It is crucial that companies looking to engage the services of an ASP forge an ironclad service-level agreement with them that includes a financial repayment in the event that the ASP fails to perform to the terms stipulated by the SLA. Sensitive to the issue of performance, the ASP Industry Consortium is working hard to develop interoperability standards and certification processes for ASPs to assuage concerns of the marketplace. For more information about ASPs, visit the ASP Industry Consortium (www.aspindustry.org), the ASP portal (www.webharbor.com), or ASP News (www.aspnews.com).

SUMMARY

IP is here to stay and is profoundly changing the nature of telecommunications at its most fundamental levels. Applications for IP range from carrier-class voice that is indistinguishable from that provided by traditional circuit-switching platforms to best-effort services that carry no service quality guarantees. The interesting thing is that all of the various capability levels made possible by the incursion of IP have an application in modern telecommunications and are being implemented at a rapid fire pace.

A tremendous amount of hype is still associated with IP services as they edge their way into the protected fiefdoms of legacy technologies, and implementers and customers alike must be wary of "brochureware" solutions and the downside of the Jurassic Park Effect, which warns that just because you *can* implement an IP telephony solution doesn't necessarily mean you *should*. Hearkening back once again to the old telephone company adage that observes "If it ain't broke, don't fix it," buyers and implementers alike must be cautious as they plan their IP migration strategies. The technology offers tremendous opportunities in consolidating services, making networks and the companies that operate them more efficient, saving costs, and passing those savings on to the customer. Ultimately, however, IP's promise lies in its ubiquity and its ability to tie services together and make them part of a unified delivery strategy. The name of the game is service, and IP provides the bridge that enables service providers to jump from a technology-centric focus to a renewed focus on the services that customers care about.

BIBLIOGRAPHY

BOOKS

Goralski, Walter and Matthew Kolon. *IP Telephony*. New York; McGraw-Hill, 2000.

Minoli, Daniel and Emma Minoli. *Delivering Voice Over IP Networks*. New York; John Wiles and Sons, 1998.

WEB RESOURCES

Network World and DigiNet Corporation. "Are You Ready for Voice Over IP?" Available at www.nwfusion.com/whitepapers/voip_wp.html.

Rendleman, John. "Voice Over Net Coming in Loud, Clear." Available at www.msnbc.com/news/296369.asp.

Walker, John Q. and Daniel McCullough. "Interested in VoIP? How to Proceed." A white paper available from Ganymede Software at www.nwfusion.com/whitepapers/ganymede/ganymede_wp.html.

ARTICLES

Bachinsky, Shmuel. "Bridging the Gap with MGCP." *Communications Systems Design*, September 1999, page 9.

Botting, Chris. "Switching Software Boosts IP Telephony." *Network World*, November 23, 1998, page 37.

Bradner, Scott. "IP Phone or Internet Phone?" *Network World,* October 4, 1999, page 56.

Brewer, John. "The IP App That's Ready for Primetime." *CTI,* May 1999, page 88.

Brothers, Art. "The Sins of IP Telephony." *America's Network,* December 1, 1998, page 54.

Brumfield, Randy. "What It Takes to Join the Carrier Class." *Internet Telephony,* May 1999, page 80.

Chappell, Laura. "Migrating to IP." *Network World,* October 18, 1999, page 63.

Connor, Chris. "The Network Service-Provider Challenge." *Communications News,* September 1999, page 90.

Cookish, Mike. "Bringing Policy to LANs and WANs Via 802.1p and DiffServ." *Communications News,* September 1999, page 56.

Cray, Andrew. "Voice over IP: Here's How." *Data Communications,* April 1998, page 44.

Davis, Brad. "The Standards Shuffle." *Telephony,* August 23, 1999, page 52.

DeVeaux, Paul. "Programming (In)flexibility." *America's Network,* September 1, 1999, page 50.

———. "The New World of IP Billing." A supplement to *America's Network,* page S6.

———. "Voice Over IP: Promise or Problems?" *America's Network,* May 1, 1999, page 61.

Doolan, Paul. "QoS." *CTI,* May 1999, page 121.

"Driving IP." *America's Network,* September 15, 1999, page 18.

Duffy, Jim. "Cisco Bringing Convergence to Small Offices." *Network World,* August 9, 1999, page 60.

———. "Cisco Unveils Packet Telephony Gear." *Network World,* August 2, 1999, page 8.

———. "Why MPLS Matters to Enterprise Networks." *Network World,* June 21, 1999, page 30.

Dzubeck, Frank. "Application Service Providers: An Old Idea Made New." *Network World,* August 23, 1999, page 49.

The editors of PriceWaterhouseCoopers' *Technology Forecast 1999.* "Internet Backbone and Service Providers: A Market Overview." *America's Network*, April 15, 1999, page 18.

Ellerin, Susan. "Network Management Platforms Make the Grade." *Network World*, September 13, 1999, page 73.

Fisher, Dave. "Defrauding the Fraudsters." *America's Network*, May 1, 1999, page 53.

Foster, Paula. "Meeting in the Cloud: Next-Gen Networks, New Enhanced Services." *CTI*, June 1999, page 82.

Friedrichs, Terra. "Talk is Cheaper with Voice Over IP." *Data Communications*, November 21, 1998, page 21.

Gately, Joe. "Road-Warrioring Made Easy." *Communications News*, September 1999, page 22.

Haramaty, Lior. "VoIP: The Opportunity. Why You Should Become an ITSP." *Internet Telephony*, July 1999, page 32.

Hindin, Eric. "Say What?" *Network World*, August 17, 1999, page 37.

"Integrated Data Platform for VoIP." *America's Network*, September 15, 1999, page 78.

"Internet Call Waiting Gets User-Friendly Upgrades." *America's Network*, April 15, 1999, page 14.

"IP/ATM Platform Tackles Brain as well as Brawn." *America's Network*, February 15, 1999, page 14.

Keegan, Paul. "Is This the Death of Packaged Software?" *Upside*, October 1999, page 139.

Klaiber, Doug. "A New Switching Architecture for a New Competitive Environment." *CTI*, June 1999, page 78.

Krapf, Eric. "Visions of the New Public Network." *Business Communications Review*, May 1999, page 20.

Kraskey, Tim and Jim McEachern. "Next-Generation Voice Services." *Network Magazine*, June 1999.

Lawrence, Jeff. "Integrating IP and SS7 Technologies." *CTI*, April 1999, page 58.

Layland, Robin. "QoS: Moving Beyond the Marketing Hype." *Data Communications*, April 21, 1999, page 17.

Levine, Shira. "The Generation Gap." *America's Network*, August 1, 1999, page 52.

———. "TMN: Dead or Alive?" *America's Network*, July 15, 1999, page 40.

Liebrecht, Don. "A TCP/IP-Based Network for the Automotive Industry." *America's Network*, May 1, 1999, page 67.

Lindstrom, Annie. "When Will Prepay in the USA Have its Day?" A supplement to *America's Network*, page S9.

Lippis, Nick. "Enterprise VoIP: Two to Go." *Data Communications*, July 1999, page 19.

"Multiprotocol Label Switching (MPLS)." From ATG's Communications & Networking Technology Guide Series, sponsored by Ennovate. Available at www.techguide.com.

Musthaler, Linda. "ASPs Make Strong Case for Renting vs. Buying Apps." *Network World*, September 27, 1999, page 37.

Newman, David. "VoIP Gateways: Voicing Doubts?" *Data Communications*, September 1999, page 71.

Noll, Michael. "Internet Tele-Phoney." *Telecommunications*, date unknown, page 42.

Nolle, Tom. "VoIP: Stalled at the Demarc." *BCR*, April 1999, page 12.

Ozur, Mark. "IP: Redefining the Telecommunications Industry." *Internet Telephony*, September 1998, page 84.

Pack, Charlie and James Gordon. "Engineering the PSTN." *America's Network*, September 15, 1999, page 16.

Pappalardo, Denise. "ASP Attack." *Network World*, September 27, 1999, page 70.

Pappalardo, Denise. "Enron Building Bandwidth the IP Way." *Network World*, August 2, 1999, page 21.

Pappalardo, Denise. "Voice Over IP Still Has Hurdles to Clear." *Network World*, September 20, 1999, page 9.

Petrosky, Mary. "Policy Capabilities Help Drive RSVP's Renaissance." *Network World*, July 5, 1999, page 33.

Pierson, Rick. "SS7—Stepping Stone to the Future." *Internet Telephony*, September 1998, page 70.

"Real-Time IP Billing." *America's Network*, September 15, 1999, page 79.

Roberts, Lawrence. "Judgment Call—Quality IP." *Data Communications*, August 1999, page 64.

Robins, Marc. "IP Breathes New Life Into the Traditional PBX Market." *Internet Telephony*, July 1999, page 22.

Robins, Marc. "Moving Beyond Plain Old VoIP." *Internet Telephony*, May 1999, page 22.

Rohde, David. "Is Qwest Losing Its IP Religion?" *Network World*, June 28, 1999, page 38.

Romascanu, Dan. "Standard Puts Keen Eye on Switched Nets." *Network World*, October 18, 1999, page 51.

Skran, Dale. H.323: "Forward, March!" *CTI*, July 1999, page 102.

Stenson, Tom. "The QoS Quagmire." *Network World*, September 6, 1999, page 53.

Stephenson, Ashley. "Controlling Oversubscription." *Communications News*, May 1999, page 60.

Taylor, Kieran. "Leveraging SS7 for Converged Voice and Data." *CTI*, September 1998, page 58.

Titch, Steven. "Mediation Devices: Shouldering the Load." *Data Communications*, NPN, page 77.

Tracey, Lenore V. "Internet Telephony: Who's Buying?" *Telecommunications*, September 1999, page 31.

"Unleash the Power: Building Multi-Service IP Networks with ATM Cores." A white paper published by the ATM Forum. Available at www.atmforum.com.

VanderBrug, Gordon. "Making QoS Work With Voice Over IP." *Communications News*, April 1999, page 56.

Wallman, Roger. "The Case for H.323 in Video Conferencing." *Internet Telephony*, June 1999, page 78.

Wallner, Paul. "The Time Is Right." *Communications News*, July 1999, page 54.

Wexler, Joanie. "QoS: What Can Service Providers Deliver?" *Business Communications Review*, April 1999, page 25.

REDEFINING THE SERVICE PROVIDER

When we started this technological journey in the introduction of the book, we discussed 16 characteristics of the evolving network and how they are affecting technology deployment and development, company positioning, and service provisioning. Let's go back and visit that list again in light of the relative roles of technology, companies, and services that we have discussed.

Customers no longer want to buy technology. This is an absolute truth that cannot be ignored by service providers if they want to continue to play an active role in their own industry. Customers no longer want to buy technology because, to a large extent, they no longer care about it. Their concern is the usage of the technology and the manner in which they will convert the technology delivered by service providers into a competitive advantage in their own markets. This represents an enormous opportunity for service providers who have the wherewithal to probe their customers' marketplaces and offer solutions based on communications technology that will help those customers position themselves competitively among their peers.

To say that customers do not want to buy technology is perhaps a bit overstated. They clearly have a need for the technology behind telecommunications and therefore want to buy it. That is not, however, where their focus lies. Technology is a means to an end in the eyes of the customer, not an end unto itself. Purveyors of technology who fail to understand that will become second-rate players and will be sidelined by those who *do* understand the difference.

Bandwidth has become a commodity. Strategic planners often talk about the SWOT Model (*Strengths, Weaknesses, Opportunities, Threats*) when analyzing themselves in the context of the markets they serve (see Figure 6-1). They typically begin by examining their own strengths, a natural place to start, and usually come up with a substantial list of abilities that make the company strong. Feeling good about themselves, they often stop the analysis at this point or only give cursory coverage to the other three areas. This is a bad practice because the exercise is valuable and should be done in its entirety for maximum benefit. Failure to analyze functional weaknesses, marketplace opportunities, or competitive threats can lead to exposure.

When service providers analyze their strengths, they typically come up with the following list:

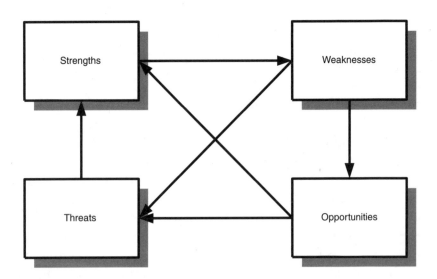

FIGURE 4-1   The SWOT model

- History and company legacy
- Reputation for good service
- Strong customer base
- In-place, capable network backbone
- People
- Size

Upon careful analysis, the list they assemble for weaknesses often includes the following:

- History and company legacy
- Less than optimal service quality and time to market for new products and services
- Declining marketshare
- Regulatory hobbles
- Large size and therefore slow-moving

Note the various similarities with the first list. Some characteristics that were seen as strengths in the legacy world are now being viewed as liabilities. This obviously points to an area that deserves attention from strategic decision-makers.

Opportunities commonly listed by service providers include these circumstances:

- The ability to sell to the competition
- The opportunity to manage the competitors' networks for them as an outsource company
- The ability to offer world-class provisioning services
- Opportunities to take advantage of new technologies in the network such as IP, wireless, and DSL to provide enhanced services
- Become an ISP and take advantage of the in-place network and its surrounding capabilities

Finally, threats commonly voiced include the following:

· A loss of marketshare to faster moving, more aggressive competitors
· Increasingly (in some eyes) punitive regulatory environments
· Growing competition
· A reduction of product profit margins

One of the greatest threats facing the incumbent service providers is the commoditization of bandwidth as it becomes available in volume and at near-zero prices. All of the concerns expressed in each of the four areas shown earlier point to the need to move from being a commodity provider to being a true service provider in a competitive marketplace.

The SWOT model can be a valuable tool for analyzing market positions and for determining strategic directions as companies move from undifferentiated commodities to specialized services. Strengths are comparatively easy to assess and, once understood, help to understand a company's weaknesses. Weaknesses are usually functions a company either does not perform at all or performs at a level that is perceived to be inferior to those the competition offers. Weaknesses in turn help to identify opportunities to either add new services or improve existing ones. Finally, threats must be managed and dealt with. They typically show themselves as competitors stealing marketshare or as internal weaknesses that give the competition some other form of advantage.

Two other relationships in the model must be mentioned. Weaknesses typically lead to opportunities, but they also help to illustrate the presence of threats. Competitors, like animals, tend to strike at their enemy's weakest point, and those are the areas that should be identified and addressed early on.

The other relationship observes that those opportunities, if identified and acted upon, help a company shore up and add to its strengths. The fact that bandwidth, formerly the greatest competitive strength of the legacy service providers, is becoming a widely traded commodity says that those providers must change the way they do business or face the classic business death spiral.

Customers are demanding more bandwidth. As bandwidth becomes cheaper and more widely available, applications are being developed that require it in larger and larger quantities. Bandwidth must therefore become one component of a full-service provider's array of services. The bandwidth must be universally accessible, cost-effective, scalable, and must provide access to a wide array of related services. Today service providers are deploying technologies that will enable them to deliver that bandwidth such as cable modems, *digital subscriber lines* (DSLs), and high-speed wireless technologies like *local multipoint distribution systems* (LMDSs).

The role of the service provider has changed dramatically. There was a time when "service provider" referred to a company that sold access and transports. Today service providers sell clouds. They have come to realize that success comes from understanding what the customer's business requirements are and offering solutions to the problems they face at competitive prices without burdening the customer with the details of the underlying technology.

Managerial structures are changing. In lockstep with the flattening of computer networks, managerial structures are flattening as well. Corporations are shifting much of the operational and tactical decision-making responsibility to lower managerial levels, thus accelerating the pace at which decisions can be made and putting responsibility for those decisions in the hands of managers best equipped to make them. At the same time, these companies maintain a philosophical hierarchy by creating a corporate vision that the peer organizations look up to for guidance. Thus, they achieve their corporate goals in ways best suited to each suborganization but are all attuned to the same vision about where the company is going.

The networks that support the evolving corporations are evolving. Again, as corporations move from being monolithic, vertically integrated organizations to virtual cluster companies, the networks they deploy internally must evolve with them. The need for flexibility, speed, and scalability has brought about the development of the *virtual private network* (VPN), a unique combination of public network infrastructure and secure transport protocols that delivers the best of the public and private network worlds.

The local loop is evolving. In keeping with customer demands for bandwidth and the service providers' need to satisfy those demands, service providers are deploying improved local loop technologies that deliver higher bandwidth without requiring massive physical upgrades to the outside plant. These include DSL, 56K modems, wireless local loop technologies, and cable modems.

Wireless access technologies are becoming centrally important. Although bandwidth is in great demand, mobility and flexibility are equally sought after. Not only is wireless infrastructure less costly to deploy, it can also be installed much faster than its wire-based relative, providing a time-to-market advantage for companies able to deploy services over it. For this reason alone, wireline companies are buying their way into the wireless domain as quickly as they can.

The fabric of the transport network is changing. As one executive recently observed, "This is not your mother's favorite circuit switch anymore." If service providers are to become all things to all applications, they must deploy a technology-rich cloud that is fully capable of delivering a wide variety of *quality of service* (QoS) levels universally at prices that customers will pay. Circuit switching will be in place for some time to come and will continue to be a central component of the world's *wide area network* (WAN) fabric. Over time, however, it will be replaced with a packet-based infrastructure and a suite of control protocols that will enable it to deliver the same quality of service that circuit switching has been known for nearly 100 years.

IP is ascending to the throne of the protocol monarchy. The ascendancy of IP is inevitable. It is a global protocol found in all network operating systems, offers a universally accepted addressing scheme, and interfaces with all network architectures. The beauty of IP is that it does not replace existing protocols, but rather provides the interstitial fabric that ties them all together, enabling service providers to unify their product and service offerings under a common, full-service cloud.

The Internet will continue to be a focal point for the evolving network-centric world. Although it may not replace other networks, it will certainly compete with them for customers as QoS protocols and improved infrastructures make it capable of offering the same services delivered today by legacy networks.

Management roles are evolving, sometimes in unusual ways. This is not trendy; it is necessary. The staid roles of traditional management hierarchies no longer work in the accelerated world of IP networking and virtual corporations. Knowledge managers, once the stuff of Ray Bradbury and Arthur C. Clarke novels, are becoming key personnel in the modern, knowledge-driven corporation.

The knowledge-based corporation is a reality, and the knowledge worker is on the rise. This is a time in the telecommunications industry for finesse, not brute force. The force of raw technology is giving way to the finesse of knowledge-managed technology, as service providers learn how to take advantage of data mining and analysis techniques to help them meet customer requirements. Leadership no longer goes to the company with the best technology, but rather to the company that has the right technology and the customer knowledge to properly apply it.

Intelligence is migrating from the center of the network out toward the user's access device. David Isenberg has written about the movement of network intelligence from the center of the service provider network to the edge of the network and beyond. Today, with the rise of IP telephony, service functions formerly performed by core network elements such as *Signaling System 7's* (SS7) *Service Controls Points* (SCPs) and *Signal Transfer Points* (STPs) are now being managed by intelligent client devices, in many cases owned and operated by the customer. Improvements in the speed and availability of specialized *digital signal processors* (DSPs) and application-specific chipsets have now made it possible to distribute the network intelligence, resulting in a faster and more efficient service creation environment.

The players in the game are changing. Here is how one *Incumbent Local Exchange Carrier* (ILEC) executive character-ized the industry today: "Imagine a playing field. On the playing field are three rugby players, a swimmer, a couple of lacrosse people, two basketball forwards, a javelin thrower, seven chess grand masters, a dart champion, and a team of skydivers. They are told to shake hands and go play. Today that's what this industry feels like a lot of the time." The characterization is a good one. This industry no longer comprises a small collection of companies with well-defined responsibilities, territories, and rules that govern them. It has become something of a free-for-all, and although that is not necessarily bad, it does inject a degree of chaos that can make life difficult for those inside the industry and confusing for those who want to interact with it.

The traditional, *local access and transport area* (LATA)-bound market is a thing of the past. An old Asian curse is "May you live in interesting times." In the past, although service providers often groused about the LATA restrictions that lim-ited their abilities to enter new markets, they clearly understood them and knew well the rules of behavior. Today all bets are off. As local companies are allowed into long-distance markets, as long-distance companies plunge into the local market, and as both buy cable properties and deploy wireless services, the rules that govern behavior in the marketplace are anyone's guess. "Be careful what you wish for—you might get it" could well be the defining mantra for today's telecommunications industry.

MOVING FORWARD

In light of all that we have discussed, what should service providers be doing to competitively position themselves? A short list describes the key steps.

Manage knowledge, not technology. Think like a customer. Become knowledgeable of the environment they operate in and create products and services that will serve their needs there. Don't try to sell them what you have; sell them what they *need*.

Sell solutions, not technologies and deliver them from an all-purpose, multifunctional network cloud.

Find a parade and get in front of it. Perform SWOT analyses of all major functional areas in the company that could be attacked by competitors and identify the weaknesses. Once found, strengthen them by either modifying internal practices or allying with strategic partners. Do it for your own company; do it for that of your customers.

Build a technology-based infrastructure that is capable, robust, and future-proof. Determine the direction that your company will take and build an infrastructure that will take you there with your customers. If analyses show that IP is the right technology, then by all means make it part of the strategic technology vision of the company. Do not, however, succumb to hype and brochureware, and heed the teachings of the Jurassic Park Effect.

Think of your company as if it were two companies: a backbone provider and a professional services provider. You may not be able to physically restructure the corporation, but you can view it as if it were restructured. The backbone company provides the necessary infrastructure, while the professional services company creates the services that will be delivered over the backbone. This imaginary structure forces a gap between the technology and the services, which is crucial; they are *not* the same. One is a vehicle; the other is the service that rides over the vehicle. Both are necessary but independent of each other.

Never forget what got your company where it is in the first place. Data used to be considered to be the killer application for voice networks because it could ride "free" atop the highly lucrative voice services. Today the opposite is becoming true. Voice is becoming the killer application for data networks because, once packetized, it can ride "for free" atop the data stream.

Build an independent knowledge management organization. Make them responsible for data mining, customer relationship management, knowledge management, and, by extension, *enterprise resource planning* (ERP). Support the need for knowledge sharing at the highest management levels in the company and make it clear that their efforts are central to the success of the corporation.

Let's face it. It's a whole new ball game out there and the customers are making the rules. The service providers, however, own the stadium, and the game is theirs to lose.

APPENDIX

This appendix is comprised of three sections. The first is a list of common industry acronyms shown in Table A-1. The second is an essay entitled "Selling to the Third Tier: Strategies for Success." The third is a list of useful URLs.

TABLE A-1   Common Industry Acronyms

AAL	ATM Adaptation Layer
AARP	AppleTalk Address Resolution Protocol
ABM	Asynchronous Balanced Mode
ABR	Available Bit Rate
AC	Alternating Current
ACD	Automatic Call Distribution
ACELP	Algebraic Code-Excited Linear Prediction
ACF	Advanced Communication Function
ACK	Acknowledgment
ACM	Address Complete Message
ACSE	Association Control Service Element
ACTLU	Activate Logical Unit
ACTPU	Activate Physical Unit
ADCCP	Advanced Data Communications Control Procedures

(Continues)

TABLE A-1   Continued

ADM	Add/Drop Multiplexer
ADPCM	Adaptive Differential Pulse Code Modulation
ADSL	Asymmetric Digital Subscriber Line
AFI	Authority and Format Identifier
AIN	Advanced Intelligent Network
AIS	Alarm Indication Signal
ALU	Arithmetic Logic Unit
AM	Administrative Module (Lucent 5ESS)
AM	Amplitude Modulation
AMI	Alternate Mark Inversion
AMP	Administrative Module Processor
AMPS	Advanced Mobile Phone System
ANI	Automatic Number Identification (SS7)
ANSI	American National Standards Institute
ANX	Automotive Network Exchange
APD	Avalanche Photodiode
API	Application Programming Interface
APPC	Advanced Program-to-Program Communication
APPN	Advanced Peer-to-Peer Networking
APS	Automatic Protection Switching
ARE	All Routes Explorer (Source Route Bridging)
ARM	Asynchronous Response Mode
ARP	Address Resolution Protocol (IETF)
ARPA	Advanced Research Projects Agency
ARPANET	Advanced Research Projects Agency Network
ARQ	Automatic Repeat Request
ASCII	American Standard Code for Information Interchange
ASI	Alternate Space Inversion
ASIC	Application Specific Integrated Circuit

ASK	Amplitude Shift Keying
ASN	Abstract Syntax Notation
ASP	Application Service Provider
AT&T	American Telephone and Telegraph
ATDM	Asynchronous Time Division Multiplexing
ATM	Asynchronous Transfer Mode
ATM	Automatic Teller Machine
ATMF	ATM Forum
ATU-C	ADSL Transmission Unit-Central Office
ATU-R	ADSL Transmission Unit-Remote
AWG	American Wire Gauge
B8ZS	Binary 8 Zero Substitution
BANCS	Bell Administrative Network Communications System
BBN	Bolt, Beranak, and Newman
BBS	Bulletin Board Service
Bc	Committed Burst Size
BCC	Blocked Calls Cleared
BCC	Block Check Character
BCD	Blocked Calls Delayed
BCDIC	Binary Coded Decimal Interchange Code
Be	Excess Burst Size
BECN	Backward Explicit Congestion Notification
BER	Bit Error Rate
BERT	Bit Error Rate Test
BGP	Border Gateway Protocol (IETF)
BIB	Backward Indicator Bit (SS7)
B-ICI	Broadband Intercarrier Interface

(Continues)

TABLE A-1   Continued

BIOS	Basic Input/Output System
BIP	Bit Interleaved Parity
B-ISDN	Broadband Integrated Services Digital Network
BISYNC	Binary Synchronous Communications Protocol
BITNET	Because It's Time Network
BITS	Building Integrated Timing Supply
BLSR	Bidirectional Line Switched Ring
BOC	Bell Operating Company
BPRZ	Bipolar Return to Zero
BRI	Basic Rate Interface
BRITE	Basic Rate Interface Transmission Equipment
BSC	Binary Synchronous Communications
BSN	Backward Sequence Number (SS7)
BSRF	Bell System Reference Frequency
BTAM	Basic Telecommunications Access Method
BUS	Broadcast Unknown Server
C/R	Command/Response
CAD	Computer-Aided Design
CAE	Computer-Aided Engineering
CAM	Computer-Aided Manufacturing
CAP	Carrierless Amplitude/Phase modulation
CAP	Competitive Access Provider
CARICOM	Caribbean Community and Common Market
CASE	Common Application Service Element
CASE	Computer-Aided Software Engineering
CAT	Computer-Aided Tomography
CATIA	Computer-Assisted Three-Dimensional Interactive Application
CATV	Community Antenna Television
CBEMA	Computer and Business Equipment Manufacturers Association

CBR	Constant Bit Rate
CBT	Computer-Based Training
CC	Cluster Controller
CCIR	International Radio Consultative Committee
CCIS	Common Channel Interoffice Signaling
CCITT	International Telegraph and Telephone Consultative Committee
CCS	Common Channel Signaling
CCS	Hundred Call Seconds per Hour
CD	Collision Detection
CD	Compact Disc
CDC	Control Data Corporation
CDMA	Code Division Multiple Access
CDPD	Cellular Digital Packet Data
CD-ROM	Compact Disc-Read-Only Memory
CDVT	Cell Delay Variation Tolerance
CEI	Comparably Efficient Interconnection
CEO	Chief Executive Officer
CEPT	Conference of European Postal and Telecommunications Administrations
CERN	European Council for Nuclear Research
CERT	Computer Emergency Response Team
CES	Circuit Emulation Service
CEV	Controlled Environmental Vault
CFO	Chief Financial Officer
CGI	Common Gateway Interface (Internet)
CHAP	Challenge Handshake Authentication Protocol
CICS	Customer Information Control System
CICS/VS	Customer Information Control System/Virtual Storage

(Continues)

TABLE A-1 Continued

CIDR	Classless Interdomain Routing (IETF)
CIDR	Classless Interdomain Routing
CIF	Cells in Frames
CIO	Chief Information Officer
CIR	Committed Information Rate
CISC	Complex Instruction Set Computer
CIX	Commercial Internet Exchange
CKO	Chief Knowledge Officer
CLASS	Custom Local Area Signaling Services (Bellcore)
CLEC	Competitive Local Exchange Carrier
CLLM	Consolidated Link Layer Management
CLNP	Connectionless Network Protocol
CLNS	Connectionless Network Service
CLP	Cell Loss Priority
CM	Communications Module (Lucent 5ESS)
CMIP	Common Management Information Protocol
CMISE	Common Management Information Service Element
CMOL	CMIP over LLC
CMOS	Complementary Metal Oxide Semiconductor
CMOT	CMIP over TCP/IP
CMP	Communications Module Processor
CNE	Certified NetWare Engineer
CNM	Customer Network Management
CNO	Chief Networking Officer
CNR	Carrier-to-Noise Ratio
CO	Central Office
CoCOM	Coordinating Committee on Export Controls
COMC	Communications Controller
CONS	Connection-Oriented Network Service

CORBA	Common Object Request Broker Architecture
COS	Class of Service (APPN)
COS	Corporation for Open Systems
CoS	Class of Service
COT	Central Office Terminal
CPE	Customer Premises Equipment
CPU	Central Processing Unit
CRC	Cyclic Redundancy Check
CRM	Customer Relationship Management
CRT	Cathode Ray Tube
CRV	Call Reference Value
CS	Convergence Sublayer
CSA	Carrier Serving Area
CSMA	Carrier Sense Multiple Access
CSMA/CA	Carrier Sense Multiple Access with Collision Avoidance
CSMA/CD	Carrier Sense Multiple Access with Collision Detection
CSU	Channel Service Unit
CTI	Computer Telephony Integration
CTIA	Cellular Telecommunications Industry Association
CTO	Chief Technology Officer
CTS	Clear to Send
CU	Control Unit
CVSD	Continuously Variable Slope Delta Modulation
D/A	Digital-to-Analog
DA	Destination Address
DAC	Dual Attachment Concentrator (FDDI)
DACS	Digital Access and Cross-Connect System
DARPA	Defense Advanced Research Projects Agency

(Continues)

TABLE A-1   Continued

DAS	Dual Attachment Station (FDDI)
DASD	Direct Access Storage Device
DBS	Direct Broadcast Satellite
DC	Direct Current
DCC	Data Communications Channel (SONET)
DCE	Data Circuit-Terminating Equipment
DCN	Data Communications Network
DCS	Digital Cross-Connect System
DCT	Discrete Cosine Transform
DDCMP	Digital Data Communications Management Protocol (DNA)
DDD	Direct Distance Dialing
DDP	Datagram Delivery Protocol
DDS	Dataphone Digital Service (Sometimes Digital Data Service)
DDS	Digital Data Service
DE	Discard Eligibility (LAPF)
DECT	Digital European Cordless Telephone
DES	Data Encryption Standard (NIST)
DHCP	Dynamic Host Configuration Protocol
DID	Direct Inward Dialing
DiffServ	Differentiated Services
DIP	Dual Inline Package
DLC	Digital Loop Carrier
DLCI	Data Link Connection Identifier
DLE	Data Link Escape
DLEC	Data Local Exchange Carrier
DLSw	Data Link Switching
DM	Delta Modulation
DM	Disconnected Mode
DMA	Direct Memory Access (computers)

DMAC	Direct Memory Access Control
DME	Distributed Management Environment
DMS	Digital Multiplex Switch
DNA	Digital Network Architecture
DNIC	Data Network Identification Code (X.121)
DNIS	Dialed Number Identification Service
DNS	Domain Name System (IETF)
DOCSIS	Data Over Cable Structured Interface Standard
DOD	Direct Outward Dialing
DOD	Department of Defense
DOJ	Department of Justice
DOV	Data over Voice
DPSK	Differential Phase Shift Keying
DQDB	Distributed Queue Dual Bus
DRAM	Dynamic Random Access Memory
DSAP	Destination Service Access Point
DSI	Digital Speech Interpolation
DSL	Digital Subscriber Line
DSLAM	Digital Subscriber Line Access Multiplexer
DSP	Digital Signal Processor
DSR	Data Set Ready
DSS	Digital Satellite System
DSS	Digital Subscriber Signaling System
DSU	Data Service Unit
DTE	Data Terminal Equipment
DTMF	Dual Tone Multifrequency
DTR	Data Terminal Ready
DWDM	Dense Wavelength Division Multiplexing

(Continues)

TABLE A-1   Continued

DXI	Data Exchange Interface
E/O	Electrical-to-Optical
EBCDIC	Extended Binary Coded Decimal Interchange Code
ECMA	European Computer Manufacturer Association
ECN	Explicit Congestion Notification
ECSA	Exchange Carriers Standards Association
EDFA	Erbium-Doped Fiber Amplifier
EDI	Electronic Data Interchange
EDIBANX	EDI Bank Alliance Network Exchange
EDIFACT	Electronic Data Interchange for Administration, Commerce, and Trade (ANSI)
EFCI	Explicit Forward Congestion Indicator/Indication
EFTA	European Free Trade Association
EGP	Exterior Gateway Protocol (IETF)
EIA	Electronics Industry Association
EIGRP	Enhanced Interior Gateway Routing Protocol
EIR	Excess Information Rate
EMBARC	E-mail Broadcast to a Roaming Computer
EMI	Electromagnetic Interference
EMS	Element Management System
EN	End Node
ENIAC	Electronic Numerical Integrator and Computer
EO	End Office
EOC	Embedded Operations Channel (SONET)
EOT	End of Transmission (BISYNC)
EPROM	Erasable Programmable Read-Only Memory
ERP	Enterprise Resource Planning
ESCON	Enterprise System Connection (IBM)
ESF	Extended Superframe Format

ESP	Enhanced Service Provider
ESS	Electronic Switching System
ETSI	European Telecommunications Standards Institute
ETX	End of Text (BISYNC)
EWOS	European Workshop for Open Systems
FACTR	Fujitsu Access and Transport System
FAQ	Frequently Asked Questions
FAT	File Allocation Table
FCS	Frame Check Sequence
FDD	Frequency Division Duplex
FDDI	Fiber Distributed Data Interface
FDM	Frequency Division Multiplexing
FDMA	Frequency Division Multiple Access
FDX	Full-Duplex
FEBE	Far End Block Error (SONET)
FEC	Forward Error Correction
FECN	Forward Explicit Congestion Notification
FEP	Front-End Processor
FERF	Far End Receive Failure (SONET)
FET	Field Effect Transistor
FEXT	Far-End Crosstalk
FHSS	Frequency Hopping Spread Spectrum
FIB	Forward Indicator Bit (SS7)
FIFO	First In, First Out
FITL	Fiber in the Loop
FM	Frequency Modulation
FRAD	Frame Relay Access Device
FRBS	Frame Relay Bearer Service

(Continues)

TABLE A-1   Continued

FSK	Frequency Shift Keying
FSN	Forward Sequence Number (SS7)
FTAM	File Transfer, Access, and Management
FTP	File Transfer Protocol (IETF)
FTTC	Fiber to the Curb
FTTH	Fiber to the Home
FUNI	Frame User-to-Network Interface
GATT	General Agreement on Tariffs and Trade
GEOS	Geosynchronous Earth Orbit Satellites
GFC	Generic Flow Control (ATM)
GFI	General Format Identifier (X.25)
GOSIP	Government Open Systems Interconnection Profile
GPS	Global Positioning System
GSM	Global System for Mobile Communications
GUI	Graphical User Interface
HDB3	High Density, Bipolar 3 (E-Carrier)
HDLC	High-Level Data Link Control
HDSL	High-Bit-Rate Digital Subscriber Line
HDTV	High Definition Television
HDX	Half-Duplex
HEC	Header Error Control (ATM)
HFC	Hybrid Fiber/Coax
HFS	Hierarchical File Storage
HITS	Headend in the Sky
HLR	Home Location Register
HSSI	High-Speed Serial Interface (ANSI)
HTML	Hypertext Markup Language
HTTP	Hypertext Transfer Protocol (IETF)
HTU	HDSL Transmission Unit

IAB	Internet Architecture Board (Formerly Internet Activities Board)
IACS	Integrated Access and Cross-Connect System
IAD	Integrated Access Device
IAM	Initial Address Message (SS7)
IANA	Internet Address Naming Authority
ICMP	Internet Control Message Protocol (IETF)
IDP	Internet Datagram Protocol
IEC	Interexchange Carrier (*see also* IXC)
IEC	International Electrotechnical Commission
IEEE	Institute of Electrical and Electronics Engineers
IETF	Internet Engineering Task Force
IFRB	International Frequency Registration Board
IGP	Interior Gateway Protocol (IETF)
IGRP	Interior Gateway Routing Protocol
ILEC	Incumbent Local ExchangeCarrier
IML	Initial Microcode Load
IMP	Interface Message Processor (ARPANET)
IMS	Information Management System
InARP	Inverse Address Resolution Protocol (IETF)
InATMARP	Inverse ATMARP
INMARSAT	International Maritime Satellite Organization
INP	Internet Nodal Processor
InterNIC	Internet Network Information Center
IntServ	Integrated Services
IP	Internet Protocol (IETF)
IPng	IP-Next Generation
IPX	Internetwork Packet Exchange (NetWare)
ISDN	Integrated Services Digital Network

(Continues)

TABLE A-1   Continued

ISO	International Organization for Standardization
ISOC	Internet Society
ISP	Internet Service Provider
ISUP	ISDN User Part (SS7)
ITFS	Instructional Television Fixed Service
ITU	International Telecommunication Union
ITU-R	International Telecommunication Union-Radio Communication Sector
IVR	Interactive Voice Response
IXC	Interexchange Carrier
JEPI	Joint Electronic Paynets Initiative
JES	Job Entry System
JIT	Just in Time
JPEG	Joint Photographic Experts Group
LAN	Local Area Network
LANE	LAN Emulation
LAP	Link Access Procedure (X.25)
LAPB	Link Access Procedure Balanced (X.25)
LAPD	Link Access Procedure for the D-Channel
LAPF	Link Access Procedure to Frame Mode Bearer Services
LAPF-Core	Core Aspects of the Link Access Procedure to Frame Mode Bearer Services
LAPM	Link Access Procedure for Modems
LAPX	Link Access Procedure Half-Duplex
LATA	Local Access and Transport Area
LCD	Liquid Crystal Display
LCGN	Logical Channel Group Number
LCM	Line Concentrator Module
LCN	Local Communications Network
LDAP	Lightweight Directory Access Protocol (X.500)

LEC	Local Exchange Carrier
LEOS	Low Earth Orbit Satellites
LI	Length Indicator
LIDB	Line Information Database
LIFO	Last In, First Out
LIS	Logical IP Subnet
LLC	Logical Link Control
LMDS	Local Multipoint Distribution System
LMI	Local Management Interface
LMOS	Loop Maintenance Operations System
LORAN	Long-Range Radio Navigation
LPC	Linear Predictive Coding
LPP	Lightweight Presentation Protocol
LRC	Longitudinal Redundancy Check (BISYNC)
LS	Link State
LSI	Large Scale Integration
LU	Line Unit
LU	Logical Unit (SNA)
MAC	Media Access Control
MAN	Metropolitan Area Network
MAP	Manufacturing Automation Protocol
MAU	Medium Attachment Unit (Ethernet)
MAU	Multistation Access Unit (Token Ring)
MD	Message Digest (MD2, MD4, MD5) (IETF)
MDF	Main Distribution Frame
MDS	Multipoint Distribution Service
MF	Multifrequency
MFJ	Modified Final Judgment

(Continues)

TABLE A-1 Continued

MGCP	Media Gateway Control Protcol
MHS	Message Handling System (X.400)
MIB	Management Information Base
MIC	Medium Interface Connector (FDDI)
MIME	Multipurpose Internet Mail Extensions (IETF)
MIPS	Millions of Instructions per Second
MIS	Management Information Systems
ML-PPP	Multilink Point-to-Point Protocol
MLT	Mechanized Loop Testing
MMDS	Multichannel, Multipoint Distribution System
MNP	Microcom Networking Protocol
MP	Multilink PPP
MPEG	Motion Picture Experts Group
MPLS	Multiprotocol Label Switching
MPOA	Multiprotocol over ATM
MRI	Magnetic Resonance Imaging
MRP	Materials Requirement Planning
MRP-II	Manufacturing Resource Planning
MSB	Most Significant Bit
MSC	Mobile Switching Center
MSO	Mobile Switching Office
MSVC	Meta-Signaling Virtual Channel
MTA	Major Trading Area
MTBF	Mean Time Between Failure
MTP	Message Transfer Part (SS7)
MTTR	Mean Time to Repair
MTU	Maximum Transmission Unit
MVS	Multiple Virtual Storage
NAFTA	North American Free Trade Agreement

NAK	Negative Acknowledgment (BISYNC, DDCMP)
NAP	Network Access Point (Internet)
NARUC	National Association of Regulatory Utility Commissioners
NASDAQ	National Association of Securities Dealers Automated Quotations
NAT	Network Address Translation
NATA	North American Telecommunications Association
NAU	Network Accessible Unit
NCP	Network Control Program
NCSA	National Center for Supercomputer Applications
NCTA	National Cable Television Association
NDIS	Network Driver Interface Specifications
NetBEUI	NetBIOS Extended User Interface
NetBIOS	Network Basic Input/Output System
NEXT	Near-End Crosstalk
NFS	Network File System (Sun)
NIC	Network Interface Card
NII	National Information Infrastructure
NIST	National Institute of Standards and Technology (formerly NBS)
NIU	Network Interface Unit
NLPID	Network Layer Protocol Identifier
NLSP	NetWare Link Services Protocol
NM	Network Module
NMC	Network Management Center
NMS	Network Management System
NMT	Nordic Mobile Telephone
NMVT	Network Management Vector Transport Protocol
NNI	Network Node Interface
NNI	Network-to-Network Interface

(Continues)

TABLE A-1   Continued

NOC	Network Operations Center
NOCC	Network Operations Control Center
NOS	Network Operating System
NPA	Numbering Plan Area
NREN	National Research and Education Network
NRZ	Non-Return to Zero
NRZI	Non-Return to Zero Inverted
NSA	National Security Agency
NSAP	Network Service Access Point
NSAPA	Network Service Access Point Address
NSF	National Science Foundation
NTSC	National Television Systems Committee
NVOD	Near Video on Demand
NZDSF	Non-Zero Dispersion Shifted Fiber
OAM	Operations, Administration, and Maintenance
OAM&P	Operations, Administration, Maintenance, and Provisioning
OC	Optical Carrier
OEM	Original Equipment Manafacturer
OIF	Optical Interoperability Forum
OMAP	Operations, Maintenance, and Administration Part (SS7)
ONA	Open Network Architecture
ONU	Optical Network Unit
OOF	Out of Frame
OS	Operating System
OSF	Open Software Foundation
OSI	Open Systems Interconnection (ISO, ITU-T)
OSI RM	Open Systems Interconnection Reference Model
OSPF	Open Shortest Path First (IETF)
OSS	Operation Support Systems

OTDR	Optical Time-Domain Reflectometer
OUI	Organizationally Unique Identifier (SNAP)
P/F	Poll/Final (HDLC)
PAD	Packet Assembler/Disassembler (X.25)
PAL	Phase Alternate Line
PAM	Pulse Amplitude Modulation
PANS	Pretty Amazing New Stuff
PBX	Private Branch Exchange
PCI	Pulse Code Modulation
PCMCIA	Personal Computer Memory Card International Association
PCN	Personal Communications Network
PCR	Peak Cell Rate
PCS	Personal Communications Services
PDA	Personal Digital Assistant
PDU	Protocol Data Unit
PIN	Positive-Intrinsic-Negative
PING	Packet Internet Groper (TCP/IP)
PLCP	Physical Layer Convergence Protocol
PLP	Packet Layer Protocol (X.25)
PM	Phase Modulation
PMD	Physical Medium Dependent (FDDI)
PNNI	Private Network Node Interface (ATM)
POP	Point of Presence
POSIT	Profiles for Open Systems Interworking Technologies
POSIX	Portable Operating System Interface for Unix
POTS	Plain Old Telephone Service
PPP	Point-to-Point Protocol (IETF)
PRC	Primary Reference Clock

(Continues)

TABLE A-1   Continued

PRI	Primary Rate Interface
PROFS	Professional Office System
PROM	Programmable Read-Only Memory
PSDN	Packet Switched Data Network
PSK	Phase Shift Keying
PSPDN	Packet Switched Public Data Network
PSTN	Public Switched Telephone Network
PTI	Payload Type Identifier (ATM)
PTT	Post, Telephone, and Telegraph
PU	Physical Unit (SNA)
PUC	Public Utility Commission
PVC	Permanent Virtual Circuit
QAM	Quadrature Amplitude Modulation
Q-bit	Qualified Data Bit (X.25)
QLLC	Qualified Logical Link Control (SNA)
QoS	Quality of Service
QPSK	Quadrature Phase Shift Keying
QPSX	Queued Packet Synchronous Exchange
R&D	Research and Development
RADSL	Rate Adaptive Digital Subscriber Line
RAID	Redundant Array of Inexpensive Disks
RAM	Random Access Memory
RARP	Reverse Address Resolution Protocol (IETF)
RAS	Remote Access Server
RBOC	Regional Bell Operating Company
RF	Radio Frequency
RFC	Request for Comments (IETF)
RFH	Remote Frame Handler (ISDN)
RFI	Radio Frequency Interference

RFP	Request for Proposal
RHC	Regional Holding Company
RIP	Routing Information Protocol (IETF)
RISC	Reduced Instruction Set Computer
RJE	Remote Job Entry
RNR	Receive Not Ready (HDLC)
ROM	Read-Only Memory
ROSE	Remote Operation Service Element
RPC	Remote Procedure Call
RR	Receive Ready (HDLC)
RSVP	Resource Reservation Setup Protocol
RT	Remote Terminal
RTS	Request to Send (EIA-232-E)
S/DMS	SONET/Digital Multiplex System
S/N	Signal-to-Noise Ratio
SAA	Systems Application Architecture (IBM)
SAAL	Signaling ATM Adaptation Layer (ATM)
SABM	Set Asynchronous Balanced Mode (HDLC)
SABME	Set Asynchronous Balanced Mode Extended (HDLC)
SAC	Single Attachment Concentrator (FDDI)
SAN	Storage Area Network
SAP	Service Access Point (Generic)
SAPI	Service Access Point Identifier (LAPD)
SAR	Segmentation and Reassembly (ATM)
SAS	Single Attachment Station (FDDI)
SASE	Specific Applications Service Element (subset of CASE, Application Layer)
SATAN	System Administrator Tool for Analyzing Networks

(Continues)

TABLE A-1   Continued

SBS	Stimulated Brouillian Scattering
SCCP	Signaling Connection Control Point (SS7)
SCP	Service Control Point (SS7)
SCSI	Small Computer Systems Interface
SCTE	Serial Clock Transmit External (EIA-232-E)
SDH	Synchronous Digital Hierarchy (ITU-T)
SDLC	Synchronous Data Link Control (IBM)
SDS	Scientific Data Systems
SECAM	Sequential Color with Memory
SF	Superframe Format (T-1)
SGML	Standard Generalized Markup Language
SGMP	Simple Gateway Management Protocol (IETF)
S-HTTP	Secure HTTP (IETF)
SIF	Signaling Information Field
SIG	Special Interest Group
SIO	Service Information Octet
SIR	Sustained Information Rate (SMDS)
SLA	Service-Level Agreement
SLIP	Serial Line Interface Protocol (IETF)
SM	Switching Module
SMAP	System Management Application Part
SMDS	Switched Multimegabit Data Service
SMP	Simple Management Protocol
SMP	Switching Module Processor
SMR	Specialized Mobile Radio
SMS	Standard Management System (SS7)
SMTP	Simple Mail Transfer Protocol (IETF)
SNA	Systems Network Architecture (IBM)
SNAP	Subnetwork Access Protocol

SNI	Subscriber Network Interface (SMDS)
SNMP	Simple Network Management Protocol (IETF)
SNP	Sequence Number Protection
SONET	Synchronous Optical Network
SPAG	Standards Promotion and Application Group
SPARC	Scalable Performance Architecture
SPE	Synchronous Payload Envelope (SONET)
SPID	Service Profile Identifier (ISDN)
SPOC	Single Point of Contact
SPX	Sequenced Packet Exchange (NetWare)
SQL	Structured Query Language
SRB	Source Route Bridging
SRS	Stimulated Raman Scattering
SRT	Source Routing Transparent
SS7	Signaling System 7
SSL	Secure Socket Layer (IETF)
SSP	Service Switching Point (SS7)
SST	Spread Spectrum Transmission
STDM	Statistical Time Division Multiplexing
STM	Synchronous Transfer Mode
STM	Synchronous Transport Module (SDH)
STP	Signal Transfer Point (SS7)
STS	Synchronous Transport Signal (SONET)
STX	Start of Text (BISYNC)
SVC	Signaling Virtual Channel (ATM)
SVC	Switched Virtual Circuit
SXS	Step-by-Step Switching
SYN	Synchronization

(Continues)

TABLE A-1   Continued

SYNTRAN	Synchronous Transmission
TA	Terminal Adapter (ISDN)
TAG	Technical Advisory Group
TASI	Time Assigned Speech Interpolation
TAXI	Transparent Asynchronous Transmitter/Receiver Interface (Physical Layer)
TCAP	Transaction Capabilities Application Part (SS7)
TCM	Time Compression Multiplexing
TCM	Trellis Coding Modulation
TCP	Transmission Control Protocol (IETF)
TDD	Time Division Duplexing
TDM	Time Division Multiplexing
TDMA	Time Division Multiple Access
TDR	Time Domain Reflectometer
TE1	Terminal Equipment Type 1 (ISDN capable)
TE2	Terminal Equipment Type 2 (non-ISDN capable)
TEI	Terminal Endpoint Identifier (LAPD)
TELRIC	Total Element Long-Run Incremental Cost
TIA	Telecommunications Industry Association
TIRKS	Trunk Integrated Record Keeping System
TL1	Transaction Language 1
TM	Terminal Multiplexer
TMS	Time-Multiplexed Switch
TOH	Transport Overhead (SONET)
TOP	Technical and Office Protocol
TOS	Type of Service (IP)
TP	Twisted Pair
TR	Token Ring
TRA	Traffic Routing Administration

TSI	Time Slot Interchange
TSLRIC	Total Service Long-Run Incremental Cost
TSO	Terminating Screening Office
TSO	Time-Sharing Option (IBM)
TSR	Terminate and Stay Resident
TSS	Telecommunication Standardization Sector (ITU)
TST	Time-Space-Time Switching
TSTS	Time-Space-Time-Space Switching
TTL	Time to Live
TUP	Telephone User Part (SS7)
UA	Unnumbered Acknowledgment (HDLC)
UART	Universal Asynchronous Receiver Transmitter
UBR	Unspecified Bit Rate (ATM)
UBR	Unspecified Bit Rate
UDI	Unrestricted Digital Information (ISDN)
UDP	User Datagram Protocol (IETF)
UDWDM	Ultra-Dense Wavelength Division Multiplexing
UHF	Ultra High Frequency
UI	Unnumbered Information (HDLC)
UNI	User-to-Network Interface (ATM)
UNMA	Unified Network Management Architecture
UPS	Uninterruptable Power Supply
UPT	Universal Personal Telecommunications
URL	Uniform Resource Locator
USART	Universal Synchronous Asynchronous Receiver Transmitter
UTC	Coordinated Universal Time
UTP	Unshielded Twisted Pair (Physical Layer)
UUCP	Unix-Unix Copy

(Continues)

TABLE A-1 Continued

VAN	Value-Added Network
VAX	Virtual Address Extension (DEC)
vBNS	Very High Speed Backbone Network Service
VBR	Variable Bit Rate (ATM)
VBR	Variable Bit Rate
VBR-NRT	Variable Bit Rate-Non-Real-Time (ATM)
VBR-nrt	Variable Bit Rate-Non-Real Time
VBR-RT	Variable Bit Rate-Real-Time (ATM)
VBR-rt	Variable Bit Rate-Real Time
VC	Virtual Channel (ATM)
VC	Virtual Circuit (PSN)
VCC	Virtual Channel Connection (ATM)
VCI	Virtual Channel Identifier (ATM)
VDSL	Very High-Speed Digital Subscriber Line or Very High-Bit-Rate Digital Subscriber Line
VERONICA	Very Easy Rodent-Oriented Netwide Index to Computerized Archives (Internet)
VGA	Variable Graphics Array
VHF	Very High Frequency
VHS	Video Home System
VINES	Virtual Networking System (Banyan)
VIP	VINES Internet Protocol
VLF	Very Low Frequency
VLR	Visitor Location Register (Wireless/GSM)
VLSI	Very Large Scale Integration
VM	Virtual Machine (IBM)
VM	Virtual Memory
VMS	Virtual Memory System (DEC)
VOD	Video on Demand
VP	Virtual Path

VPC	Virtual Path Connection
VPI	Virtual Path Identifier
VPN	Virtual Private Network
VR	Virtual Reality
VSAT	Very Small Aperture Terminal
VSB	Vestigial Sideband
VSELP	Vector-Sum Excited Linear Prediction
VT	Virtual Tributary
VTAM	Virtual Telecommunications Access Method (SNA)
VTOA	Voice and Telephony over ATM
VTP	Virtual Terminal Protocol (ISO)
WACK	Wait Acknowledgment (BISYNC)
WACS	Wireless Access Communications System
WAIS	Wide Area Information Server (IETF)
WAN	Wide Area Network
WARC	World Administrative Radio Conference
WATS	Wide Area Telecommunications Service
W-CDMA	Wideband CDMA
WDM	Wavelength Division Multiplexing
WIN	Wireless In-building Network
WTO	World Trade Organization
WYSIWYG	What You See Is What You Get
xDSL	x-Type Digital Subscriber Line
XID	Exchange Identification (HDLC)
XNS	Xerox Network Systems
ZBTSI	Zero Byte Time Slot Interchange
ZCS	Zero Code Suppression

(Continues)

SELLING TO THE THIRD TIER: STRATEGIES FOR SUCCESS

Marketplace success today demands that would-be service or equipment providers "sell to the third tier." This refers to the fact that they must look beyond their immediate customers (the so-called second tier) and position products and services for the needs of their *customers'* customers as well. For example, a manufacturer of routers must certainly sell to their primary audience (service providers, ISPs, corporate accounts), but if they want to maintain a favorable competitive edge over the competition, they should consider the requirements of the segment of the market that constitutes the customers of their immediate marketplace as well. This requires careful strategic planning. Most service providers know their own customers' requirements well but precious little about the companies and industries that their own customers are attempting to satisfy. Although most companies understand the driving forces behind their own customers' behaviors, they generally know far less about the issues that the third tier faces. It is critical therefore that service providers and manufacturers alike learn to sell to that third tier of customers because doing so will position them optimally before their primary customers. By using data mining and knowledge management techniques to better understand the requirements of the third tier, companies can anticipate how the service provider or manufacturer can not only have the right technology in place, but actually anticipate the requirements of their own customers, thus giving them the ability to satisfy the demands of their own customers far more quickly and efficiently. This is customer service at its finest.

This approach to market analysis represents something of a departure for many businesses, yet it is a key component of success in today's rollicking competitive telecommunications marketplace. The approach can be broken into three major activity areas: a market analysis, an analysis of the impact of disruptive technologies, and a set of special considerations for businesses that operate in the multinational sphere.

At the market analysis level, service providers and manufacturers must consider the following:

- *What is the customer mix?* Within the defined markets of the second- and third-tier customers, what does the customer profile look like? What is the relative percentage of residence and business lines installed? Furthermore, are these numbers expected to change over time or is the area relatively static? What impact will phenomena such as telecommuting and the *Small Office/Home Office* (SOHO) have on these percentages?

- *Who are the primary customers?* Are they small businesses, large office-based firms, or manufacturing concerns? Or are they primarily residence customers?

- *What is the strategic growth plan for this particular customer?* Do they intend to stay small or will they grow over time? Have they announced intentions to or is their business predisposed to form strategic alliances with other companies to gain greater marketshare that might change the way you approach them?

- *To what degree is the market fragmented by the presence of multiple service providers or multiple companies in the same business as this customer?* Would providing service to this customer endanger your ability to sell the same or similar services to other potentially more lucrative or stable customers?

- *What is the nature of the regulatory environment in which this customer must be served?* Is it likely that the customer will demand services that are either not available or prohibitively expensive to provide? Does the customer have multiple locations scattered over a wide area, such that a local service provider would be unable to serve their corporate-wide needs? Furthermore, is the company a multinational, in which case their requirements for company-wide service would become even *more* complex?

- *What is the current and projected service mix of traffic that the customer will be generating?* In most markets, data represents the highest percentage of traffic volume yet generates a comparatively tiny percentage of corporate revenue. Voice, on the other hand, typically generates an enormous amount of revenue yet represents a small percentage of total transported traffic. In some markets, videoconferencing, *Voice over IP* (VoIP), Internet voice, high-resolution imaging, and other services are placing growing demands on network bandwidth reserves. What considerations must be taken into account if this is the case with the customer or market in question?

In terms of the impact of disruptive technologies, the following must be considered:

- *Has cable made a significant incursion in this market?* If so, to what degree has it been accepted by the local marketplace? Is the acceptance primarily by data services for residential customers or is there considerable business uptake as well? So far, the bulk of cable modem sales have been in the residence segments, but activity exists in the business market as well.
- *Is VoIP a serious consideration yet or is it still in the early adopter stage?* Are there indications that alternative services such as VoIP or Internet voice will receive favorable treatment in the eyes of this market? Are there "visible early adopters" who could skew the market in favor of non-traditional services?
- *To what degree have wireless access and transport technologies succeeded in this market?* Are there technological, geographical, economic, or logistical reasons to be wary of wireless as an alternative competitive solution? If so, what are they, and how can they be addressed by your own solutions?
- *Do legacy installations of technology have a toehold that must be reckoned with?* In many cases, legacy technology is

not a negative factor, but if there is a need to supplant a legacy installed base, its very presence and the fact that it has put down roots can severely impact a service provider's or manufacturer's ability to convince customers of the need for change.

- *How influential is the Internet in this marketplace?* Is it pervasively deployed, widely accepted, and heavily utilized? How many ISPs are present in this market that drive the demand for Internet services?

For multinational managers, a number of related considerations surface:

- *To what degree do liberalization and privatization in this country/market affect the ability to do business?* If the market is currently a monopoly or is severely government-controlled, how will the ability to do business change when the market is opened to full-blown competition? What will the government's role be in the newly liberalized operating environment?
- *Related to the previous questions, what is the current political environment in-country?* Does the current regime look favorably upon the presence of foreign-owned business, or is it more hostile and predisposed to do business with domestic concerns?
- *What is the availability of critical technology in-country?* Given that a certain level of infrastructure support is required for the successful deployment of new equipment and services, does a paucity of local infrastructures adversely affect the ability to deploy successfully? Furthermore, are there any technology transfer limitations that would adversely affect in-country deployment plans? Most of these restrictions are slowly disappearing but are still considerations in some countries. Consult Department of State and Commerce publications to be sure.
- *What is the in-country philosophy with regard to the Internet?* Some countries view it as a facilitator of needed

change and as a "jump-start" technology, while others view it as a socially disruptive force that is to be eschewed at all costs. Knowledge of local sentiments can help with market penetration plans.

• *To what degree do preexisting alliances, relationships, and/or technology biases exist in the market under scrutiny?* In many cases, new entrants will be faced with the very real possibility that incumbent players already have a leg up on the competition because of a preexisting local presence, familiarity with the local market, and relationships with key market players.

• *Is technology snobbery a reality in this market?* In some cases, preferred vendors or technologies are stubbornly perceived as being superior to all others, often because it requires effort and money to consider others. In these cases, it may be worthwhile to prepare a hands-on "road show" to demonstrate the characteristics and advantages of new products and services.

• *Are there any geographical issues that must be considered?* These would include the local terrain over which the infrastructure must be deployed (swamp, desert, mountainous terrain, off-limits areas), access to coastlines for transoceanic cable connectivity, vegetation density that might affect the ability to deploy line-of-sight wireless solutions, and relationships with neighboring countries that could affect the ability to deploy wide area technologies destined to connect with the rest of the world.

• *Are there any overwhelming financial issues that must be considered?* These might include the ability to repatriate profits easily, local payment-for-services considerations, the potential for force majeur, and any number of others, tangible or otherwise.

These issues may or may not be considerations in any given market, but all should be considered nonetheless. Those that are not considerations can be dismissed, but the very act of considering them may identify other issues that would not otherwise have been considered.

DEVICE ASSESSMENT CRITERIA

Manufacturers and service providers are often faced with the task of identifying the advantages and disadvantages that exist between competing devices. Based on the specific requirements of applications, services, standards mandates, or other reasons, sales and technical support personnel must often perform a comparative analysis of multiple devices. When performing such an analysis, the following should be considered:

- *Backbone versus small business*   Under this category, a number of characteristics become centrally important including
  - The number of *packets per second* (PPS) that the device can process
  - Varied protocol support (support for BGP is critical, especially the Cisco version). The problem is that the RFCs enable significant interpretation latitude.
  - The degree of QoS implementation. Does the device support proprietary QoS techniques as well as those based on standards? Does the device support DiffServ/RSVP and other techniques?
- *Interface support*   What are the interfaces that this device is capable of supporting? At a minimum, it should probably support some or all of the following:
  - T1/E1
  - DS3/E3
  - SONET/SDH
  - Line/trunk

  Additionally, the device should adequately address the following processing concerns:
  - Does the device support *input/output* (I/O) on the same board or does it manage I/O across the backplane?
  - Does the device support onboard protocol conversions?
  - Does the device have a blocking or non-blocking architecture?

What user interfaces does the device support? Cisco's is the model for the industry and has been widely imitated because of its ease of use and functionality. Others do exist, however, and should have some or all of the following characteristics:

- GUI command line interface
- In-band vs. out-of-band management messaging

- Here are some additional miscellaneous characteristics:
  - Is the device NEBS-compliant?
  - How scalable is it?
  - How strong is its survivability/fault management capability?
  - Does the device support CTI integration?
  - What kind of protocol support does it offer? Does it offer support for ATM, FR, IP, H.320/323, and so on?
  - What traffic management schemes are supported?

SUMMARY

Remember that the key to success is to use data mining, knowledge management, and customer relationship management to anticipate customer requirements and be ready to respond when they come looking for solutions, or in some cases to proactively offer solutions before they ask. Anticipatory homework that helps the service provider or manufacturer respond to both the second and third tiers will provide a significant advantage over those who stop at tier one.

CONVERGENCE ONLINE RESOURCES

The online resources listed here provide good starting points for readers wanting to delve deeper into the companies and activities behind the convergence scene.

Readers should be warned that the fluid nature of the World Wide Web may cause some of these sites to become inactive or unavailable over time. If you find a problem, please notify us so that we can post the change on the Convergence Web site.

Many thanks to Gary Kessler for tracking and providing much of this material.

TECHNOLOGY CONVERGENCE
xDSL

ADSL Forum	www.adsl.com
Analog Devices	www.analog.com/publications/whitepapers/products/whitepaper_html/content.html
AT&T Research	www.research.att.com
Dan Kegel's ADSL Page	alumni.caltech.edu/~dank/isdn/adsl.html
Paradyne's DSL Tutorial	www.alliancedatacom.com/paradyne-dsl-tutorial.htm
WPI's ADSL Page	bugs.wpi.edu:8080/EE535/hwk97/hwk3cd97/mrosner/mrosner.html
xDSL FAQs	www.telechoice.com/xdslnewz homepage. interaccess.com/~jkristof/xdsl-faq.txt

(Continues)

ATM

ABR Service Overview ·   www.atmforum.com/atmforum/library/53bytes/
backissues/others/53bytes-1095-2.html

ANSI Draft T1S1.5/96-186   ftp://ftp.t1.org/pub/t1s1/t1s1.5/
(AAL-CU, the New AAL2)   6s151860.doc

Arequipa (Application Requested   icawww.epfl.ch
Quality of Service over ATM)

ATM Address Format Papers   www.ed.ac.uk/~george/ukac-addr.html

ATM at Work, ATM Trials   www.atm.at-work.com/success_folder/
in Various Industries   Successatm.html

ATM Chip Web and Link   www.infotech.tu-chemnitz.de/~paetz/atm
Web Pages

ATM Forum   www.atmforum.com

ATM Interoperability Lab   www.iol.unh.edu
(IOL), U. NH

ATM over ADSL   www.alliancedatacom.com/
paradyne-dsl-tutorial.htm

AT&T's WWW Server   www.research.att.com

Bellcore ATM Docs (ftp)   ftp://thumper.bellcore.com/pub

Cambridge University ATM   www.cl.cam.ac.uk/Research/SRG/
Document Collection   bluebook.html

Cisco Frame Relay to ATM   www.alliancedatacom.com/
Tutorial (Includes Management,   cisco-frame-relay-atm-tutorial.htm
OAM Cells, and OAM Cell
Support)

David Blight's Telecommunications www.ee.umanitoba.ca/~blight/telecom.html
Reference Page

Fore Systems ATM ftp Archive   ftp://ftp.fore.com/pub

Herwig's ATM Page   www.icg.tu-graz.ac.at/herwig/Research/
(Lots of Pointers)   ATM/ATM.html

J.S. Turner "Gigabit ATM   www.arl.wustl.edu/~jst/gigatech/
Networking Technology" Paper,   gigatech.html
Washington University

Keshav's ATM Papers at AT&T   www.cs.att.com/csrc/keshav/papers.html

Native ATM Tutorial   www.alliancedatacom.com/
native-atm-tutorial.htm

Newbridge VIVID ATM Web Page	www.vivid.newbridge.com
North Carolina State U. ftp Archive (ATM/QoS/Multimedia Papers)	ftp://ftp.csc.ncsu.edu/pub/rtcomm
Papers on Voice-Over ATM, AAL2, and More (GDC)	www.voiceofatm.com
Purdue University ATM Info	www.cs.purdue.edu/acm/acm.useful.html
Raj Jain's ATM Info	www.cis.ohio-state.edu/~jain
Random Early Discard (RED) Paper from Floyd & Jacobson (LBL)	www-nrg.ee.lbl.gov/floyd/red.html
Telecom Finland ATM ftp Archive (ITU-T, ANSI, and Other ATM Specs.)	ftp://ftp.tele.fi/atm
White Paper about IBM's Implementation of the P-NNI	www.networking.ibm.com/pnni/pnniwp.html

VOICE, VIDEO, AND MULTIMEDIA

Articles about MPEG over ATM	www.mpeg.org/~tristan/MPEG/ starting-points.htm#atm
Multimedia over ATM Information from GDC	www.alliancedatacom.com/ gdc-multimedia-video-systems.htm
Video on Demand (VOD)	www.sntc.com/papers/netshow_wp/ netshow_wp.htm
Voice Telephony over ATM (VTOA)	cio-europe.cisco.com/public/rfc/ATM/ atm-forum/vtoa/

ATM SIGNALING

GN's Tutorial	www.webproforum.com/gnnettest/topic03.html
Nortel's Paper	www.webproforum.com/nortel2/index.html

(Continues)

Roxen's ATM Signaling Web Page   `www.roxen.com/rfc/rfc1755.html`

Trillium's Site   `www.trillium.com/whats-new/wp_atmsig.html`

PERFORMANCE

ATM Performance Measurement: Throughput, Bottlenecks, and Technology Barriers   `isdn.ncsl.nist.gov/sources/atmperf/atmperf.html`

LLNL ATM Performance Data, Cell Pacing   `www.llnl.gov/bagnet/perf-llnl-lcl-cell-pacing.html`

Network Performance (ATM)   `www.cup.hp.com/netperf/NetperfPage.html`

MPOA

MPOA.com   `www.mpoa.com`

Tech Guide   `www.techguide.com`

IP OVER ATM

IP over ATM Information   `www.stl.nps.navy.mil/~iirg/atm/ip-atm/`

More IP over ATM Info   `www.csl.sony.co.jp/person/demizu/ipnet/mlr.html`

TCP and ATM, ATM Forum (Sun)   `ftp://playground.sun.com/pub/tcp_atm`

TCP/IP Performance over ATM   `www.tisl.ukans.edu/Workshops/ATM_Performance`

IP SWITCHING

ARIS Specification (IBM)   `www.networking.ibm.com/isr/arisspec.html`

Cisco (Tag Switching)   `www.cisco.com/warp/public/732/tag/`

IETF's Multiprotocol Label Switching (mpls) Working Group   `www.ietf.org/html.charters/mpls-charter.html`

Ipsilon   `www.ipsilon.com`

| MPLS (MultiProtocol Label Switching) Site | `dcn.soongsil.ac.kr/~jinsuh/home-mpls.html` |

ATM WAN SWITCHES

Data Communications Magazine ATM Switch Stress Test	`www.data.com/Lab_Tests/ATM_Stress_Test.html`
Dataquest	`www.dataquest.com/`
IDC	`www.idcresearch.com/`
LAN Times	`www.lantimes.com/98/98feb/pcatmswi.html`
SNCI's Switch Information	`www.snci.com`
Vertical Systems	`www.verticalsystems.com/`

CABLE TELEVISION

Adelphia	`www.adelphia.com/`
Cable Labs (U.S.)	`www.cablelabs.com`
Cable Telephony, Cable Modems, and Data Services	`www.catv.org`
Cable TV Resources on the Net	`www.internic.net/nii`
Cox Cable	`www.cox.com`
@Home	`www.home.com`
IEEE 802.14 (LANs over CATV)	`www.com21.com/pages/ieee-802.14.html`
Inside Cable Magazine (U.K.)	`scitsc.wlv.ac.uk/university/sles/sm/incable.html`
National Cable Television Association (NCTA)	`www.ncta.com/home.html`

COMPUTERS AND COMPUTING

| *Computer Magazine* Archive, ZDNet's Database of 200,000 Computer-Related Articles | `cma.zdnet.com` |

(Continues)

Computer Services Benchmarking   `www.jncs.com/fpages/benchmark.htm`

Information Systems Meta-List   `www.cait.wustl.edu/cait/infosys.html`
with Basic Information about
Many Networking and Computer
Topics

The Rapidly Changing Face of   `www.digital.com/rcfoc/home.htm`
Computing Technology Journal

GENERAL TECHNICAL RESOURCES

CNET   `www.cnet.com`

Developer.com   `www.developer.com`

Open Source Initiative   `www.opensource.org`

O'Reilly and Associates   `www.ora.com`

O'Reilly Open Source Center   `opensource.oreilly.com`

Whatis.com   `www.whatis.com`

ZDNET   `www.zdnet.com`

FRAME RELAY

Act Systems (Includes   `www.acti.com`
Voice-Over FR Info)

Cisco Frame Relay to ATM   `www.alliancedatacom.com/`
Tutorial (Includes Management, `cisco-frame-relay-atm-tutorial.htm`
OAM Cells, and OAM Cell
Support)

Frame Relay Forum Archive   `www.frforum.com`
and IAs

Frame Relay Forum Technical   `ftp://ibmstandards.raleigh.ibm.com/pub/`
Committee ftp   `standards/FrameRelay`

Frame Relay Resource Center   `www.alliancedatacom.com`

Frame Relay Resources Info   `www.mot.com/MIMS/ISG/tech/frame-relay/`
(Motorola), a Great Starting Point `resources.html`

ISC Frame Relay Products   `www.infoanalytic.com/isc`

More FR Pointers (Planet)   `www.planet.net/icxc/links.html`

RadCom Voice-Over Frame Relay	`www.rad.com/networks/1995/fram-rel/` `voice.htm`
UniSPAN Consortium of Frame Relay Carriers	`www.unispan.com/`

ISDN/SS7

Ascend SS7 Information	`www.alliancedatacom.com/` `ascend-signaling-gateway-ss7.htm`
Bandwidth On Demand INteroperability Group (BONDING)	`www.hep.net/ftp/networks/bonding/`
Bellcore's ftp Site, Containing NIUF information, ISDN Service Contacts at the Bellcore Client Companies, Deployment Information, National ISDN Information, and More	`ftp://info.bellcore.com/pub/ISDN`
Bellcore's National ISDN Page and NIUF Catalog	`www.bellcore.com/ISDN/ISDN.html`
California ISDN Users Group	`www.ciug.org/`
Dan Kegel's ISDN Page, Possibly the Best Starting Point for Service, Products, and Other ISDN Information	`alumni.caltech.edu/~dank/isdn/`
Dave Hawley's isoEthernet (IEEE 802.9a) Page	`members.aol.com/dhawley/isoenet.html`
Hotlinks to ISDN Equipment Vendors	`www.primenet.com/~towens/ISDN/isdn.htm`
ISDN Systems Corp.	`www.infoanalytic.com/isc/index.html`
MicroLegend's SS7 Tutorial, an excellent tutorial	`www.microlegend.com/aboutss7.htm`
Microsoft ISDN Information	`www.microsoft.com`
MIT's ISDN Frequently Asked Questions (FAQ) Lists	`ftp://rtfm.mit.edu/pub/usenet/` `news.answers/isdn-faq/`

(Continues)

Motorola ISDN Systems Group	www.mot.com/MIMS/ISG/
National ISDN Council (Bellcore)	www.bellcore.com/NIC/
National ISDN Registry of Customer Equipment (Bellcore)	www.bellcore.com/ISDN/web_list.htm
North American ISDN Users' Forum (NIUF)	www.ocn.com/ocn/niuf/niuf_top.html
North American ISDN Users' Forum (NIUF) Catalog, IOCs, and so on	www.niuf.nist.gov/misc/niuf.html
Open Communication Networks' ISDN InfoCentre	isdn.ocn.com/
Sven De Kerpel's ISDN Information Base	igwe1.vub.ac.be/~svendk/ isdn_homepage.html
Texas ISDN Users Group	www.tiug.org/

LOCAL AREA NETWORKS (LANs)

Dave Hawley's isoEthernet (IEEE 802.9a) Page	members.aol.com/dhawley/isoenet.html
Ethernet Home Page	wwwhost.ots.utexas.edu/ethernet/ ethernet-home.html
Gigabit Ethernet Alliance	www.gigabit-ethernet.org
HSSI Specification (Cisco)	ftp://ftp.uu.net/networking/cisco/ hssi.not.Z
IEEE 802.3z Gigabit Ethernet Page	www.wco.com/~schelto/Ethernet/gigabit.htm
IEEE Higher Speed Study Group (10-Gbps Ethernet)	grouper.ieee.org/groups/802/3/10G_study/ public/
IsoEthernet (802.9a) Home Page	www.nsc.com/isoethernet/index.html
ISO/IEC Drafts for Cat 6 and Cat 7 Cable	www.iso.ch/welcome.html
LAN Acquisitions, "Who Bought Who?"	web.syr.edu/~jmwobus/compages/comfaqs/ lan-acquisitions.html
LANs over CATV (IEEE 802.14)	www.com21.com/pages/ieee-802.14.html
LBL's FDDI Document Repository	ftp://fddi.lbl.gov

| Siemon Co. Papers on 155-Mbps ATM over Cat 5 | www.siemon.com/references/techhome.html |
| U. NH Interoperability Lab (IOL): ATM, FDDI, IEEE 802.11, Fast Ethernet, Fibre Channel, IP/Routing, Network Management, Token Ring, VG-AnyLAN | www.iol.unh.edu |

NETWORK MANAGEMENT

Data Communications' Network and Systems Management Directory	www.data.com/directory/index.html#8
Network Management	smurfland.cit.buffalo.edu/NetMan/index.html
Network Management (U. of Twente)	www.nic.utwente.nl/
Network Management Forum (NMF)	www.nmf.org
SNMPinfo, Lots of SNMP, SMI, and MIB Information	www.snmpinfo.com
SNMP MIB Status	www.rampages.onramp.net/~cwk/mibs.html

NETWORKING PUBLICATIONS

BYTE Magazine	www.byte.com/
HotWired	www.hotwired.com
InfoWeek	techweb.cmp.com/iwk
LAN Magazine	www.lanmag.com
Network World	www.nww.com
Network World Fusion Web	www.nwfusion.com
PC Magazine	www.zdnet.com/pcmag

(Continues)

NETWORKING TOPICS (GENERAL)

A Series of Technical Guides about a Variety of Topics	www.techguide.com
BitPipe, the Latest on Technology Research	www.bitpipe.com
Cisco's Internetwork Design Guide	www.alliancedatacom.com/ cisco-internetwork-design.htm
Data Communications Magazine's Network Technologies Page	www.data.com/cgi-bin/IWatchIndex
HP's Network Performance Page	www.cup.hp.com/netperf/NetperfPage.html
Information Systems Meta-List, Basic Information about Many Networking and Computer Topics	www.cait.wustl.edu/cait/infosys.html
Netdictionary	www.netdictionary.com
Reference Guide for Data Communications and Networks	www.techedcon.com/refguide/ Referencepage.htm
Silicon Graphics (SGI) Technology Overview	www.sgi.com/Technology/extreme_tech.html
U. NH Interoperability Lab (IOL): ATM, FDDI, IEEE 802.11, Fast Ethernet, Fibre Channel, IP/Routing, Network Management, Token Ring, VG-AnyLAN	www.iol.unh.edu

PROTOCOLS

C.M. Heard's CRC ftp Site	ftp://ftp.vvnet.com/aal5_crc32/
The CRC Pitstop	www.ross.net/crc/index.html
Network Analysis Institute References	www.netanalysis.org/references.html
Protocol Directory, Including Frame Decodes and Technical Papers (Radcom)	www.protocols.com
Radcom Academy, an Excellent Source of Information about AppleTalk, TCP/IP, ATM, Xerox, and Other Protocols	www.radcom-inc.com/acad/protocols.htm

MODEMS AND HIGH-SPEED CARRIERS

56K.COM, an Overview of   www.56k.com
56-kbps Modem Technology

Another High Bandwidth Page   www.speciality.com/hiband/

The Telecom Corner, with Info   www.tbi.net/~jhall/
on Digital Carriers (T1, E1, and
so on) and Digital Voice, History,
Novell, Muxing, and More

CABLING

Cables and Pinouts   www.rs232.com

Cabling FAQ (LAN/WAN/   www.cis.ohio-state.edu/hypertext/faq/
ISDN Cabling Pinouts and   usenet/LANs/cabling-faq/
Standards)

Data Communications   www.alfasys.be/data.htm
Cabling FAQ

International Cable Federation   www.icf.at/main.html

Network Cabling Resources   www.oceanwave.com/technical-resources/
                                 network-admin/cabling.html

Structured Cabling System   www.andrewboon.com.au/BOONSCS1.htm
Starter Guide

Teldor Wires and Cables Links   www.elron.net/teldor/
to LAN-Related Sites, FAQs,
Discussion Lists and "Copper
Cabling Systems for High
Data Rates"

INTERNETWORKING

IEEE 802.5 Web Site   p8025.york.microvitec.co.uk

High-Speed Token Ring Alliance   www.hstra.com

Technology Overview, Terms   www.geocities.com/CollegePark/4087
and Acronyms, Design Guide,
Troubleshooting, Case Studies

(Continues)

TECHNOLOGY & NETWORK SPECIFICATIONS

All-Optical Networking Consortium	www.11.mit.edu/aon
Multiwavelength Optical Networking Consortium (MONET)Networking Consortium (MONET)	www.bell-labs.com/project/MONET
Optical Internetworking Forum (Including IP Over Fiber)	www.oiforum.com
Optical Fiber Information	www.ctr.columbia.edu/~georgios/ lightwave.html

SONET/SDH/DWDM

Cisco 12000 IP-over-SONET Information	www.cosco.com/warp/public/733/12000/ gspos_an.htm
Communications Industry Researchers (CIR)	www.cir-inc.com
DWDM Overview (Ciena)	www.techguide.com/comm/sec/dwave.pdf
DWDM Tutorial (Lucent)	www.webproforum.com/lucent3/index.html
"Intro to SONET" by J.S. Mahal & D. Moore	www.geocities.com/CapeCanaveral/Lab/ 6800/P97_JDSR.HTM
IP-over SONET Versus IP-over-ATM (Trillium)	www.trillium.com/whats-new/wp_ip.html
Oxygen, Undersea Fiber Loops	www.oxygen.org
Packet over SONET Overview	www.ncne.nlanr.net/news/workshop/ 981101/Talks/Dunn/index.htm
SONET Home Page	www.sonet.com/
SONET Interoperability Forum	www.atis.org/atis/sif/sifhom.htm

FIBRE CHANNEL

Fibre Channel Association	www.fibrechannel.com
Fibre Channel ftp (Sun)	ftp://playground.sun.com/pub/fibrechannel
Fibre Channel Loop Community	www.fcloop.org/
Fibre Channel Pointers (CERN)	www.cern.ch/HSI/fcs

PROFESSIONAL SOCIETIES AND INDUSTRY ORGANIZATIONS

ACM Special Interest Group
for Data Comm., With
Computer Comm. Reviews

`gatekeeper.dec.com/.b/doc/sigcomm/ccr/`
`overview.html`

ACM Web Site

`info.acm.org`

The Bootstrap Institute
(Doug Englebarth)

`www.bootstrap.org`

IEEE Computer Society

`www.computer.org`

Institute of Electical and
Electronic Engineers (IEEE)

`www.ieee.org`

International Communications
Association

`icanet.com/`

STANDARDS BODIES AND RELATED ORGANIZATIONS

ADSL Forum

`www.adsl.com`

Alliance for Telecommunications
Industry Solutions (ATIS),
Secretariat for ANSI T1
Committee (formerly Exchange
Carriers Standards Association
[ECSA])

`www.atis.org`

American National Standards
Institute (ANSI)

`www.ansi.org`

APPN Implementors Workshop
(AIW)

`www.raleigh.ibm.com/app/aiwhome.htm`

ATM Forum

`www.atmforum.com`

Cellular Telecommunications
Industry Association
(Really www.wow-com,
The World of Wireless)

`www.wow-com.com/consumer`

Desktop Management Task Force
(DMTF)

`www.dmtf.org`

Electronic Industries Alliance
(né Association) (EIA)

`www.eia.org`

(Continues)

European Telecommunications Standards Institute (ETSI)	`www.etsi.fr`
Excellent Set of Standards Pointers	`www.cmpcmm.com/cc/standards.html`
Frame Relay Forum	`www.frforum.com`
Gigabit Ethernet Alliance (GEA)	`www.gigabit-ethernet.org`
High-Speed Token Ring Alliance	`www.hstra.com`
HIPPI Networking Forum	`www.nets.com/hippiforum`
IEEE Standards Office	`stdsbbs.ieee.org`
Institute of Electrical and Electronics Engineers (IEEE)	`www.ieee.org`
International Electrotechnical Commission (IEC)	`www.iec.ch/home-e.htm`
International Organization for Standardization (ISO)	`www.iso.ch`
International Telecommunication Union (ITU)	`www.itu.int`
Internet Assigned Number Authority (IANA)	`www.iana.org`
Internet Engineering Task Force (IETF)	`www.ietf.org`
Internet Society (ISOC)	`www.isoc.org`
Moving Picture Experts Group (MPEG) (ISO/IEC JTC1/SC29 WG11)	`drogo.cselt.stet.it/mpeg`
Network Management Forum (NMF)	`www.nmf.org`
North American ISDN Users' Forum (NIUF)	`www.niuf.nist.gov/misc/niuf.html`
Object Management Group (OMG)	`www.omg.org`
The Open Group	`www.opengroup.org`
Optical Internetworking Forum	`www.oiforum.com/`
Public OUI Assignments	`standards.ieee.org/regauth/oui/index.html`

Society of Automotive Engineers   www.sae.org

Society of Motion Pictures and   www.smpte.org
Television Engineers (SMPTE)

SONET Interoperability Forum   www.atis.org/atis/sif/sifhom.htm

Telcordia (Formerly Bellcore)   www.telcordia.com/products/
                                 information_products.html

Telecommunications Industry   www.tiaonline.org
Association (TIA)

United States Telephone   www.usta.org
Association (USTA)

Wireless Application Protocol   www.wapforum.org
(WAP) Forum

World Wide Web Consortium   www.w3c.org
(W3C)

TECHNOLOGY AND SOCIETY

Rand Corp's "Universal Access   www.rand.org
to E-Mail: Feasibility & Social
Implications"

Technical, Economic, Public   www.spp.umich.edu/telecom/
Policy, and Social Aspects of   telecom-info.html
Telecom from University
of Michigan Covering GII,
Internet Telephony, and
Electronic Commerce

TELECOMMUNICATIONS (GENERAL INFORMATION)

Boadband Telephony Buyer's   www.indra.com/unicom
Guide

Canada Information   ai.iit.nrc.ca:80/superhighway.html
Superhighway

Canadian Telecom Market   www.angustel.ca

(Continues)

Columbia Center for Telecom   www.ctr.columbia.edu
Research, Covers a Broad Range
of Topics about Mass Media,
Telecom, and Communication in
Cyberspace, Including the Virtual
Institute of Information

David Blight's Telecommunications www.ee.umanitoba.ca/~blight/telecom.html
Reference Page

Internet Tariff Library   www.tariffnet.com

Lawrence Livermore   www-atp.llnl.gov/atp/telecom.html
National Lab

Lili Goleniewski's Telecom   www.telecomwebcentral.com
WebCentral

TELCO Exchange, Covering Digital   www.telcoexchange.com
Pricing and ISP Searching

Telecom Information Resources   www.idt.unit.no/~csurgay/
on the Internet   telecom-info.html

University of Michigan   china.si.umich.edu/telecom/
Telecom Page   telecom-info.html

TELECOMMUNICATIONS REGULATION AND POLICY

Alliance for Public Technology   apt.org/apt.html

Business Software Alliance   www.bsa.org

Center for Democracy and   www.cdt.org
Technology

The Center for Media Education   www.access.digex.net/~cme/bill.html

Common Carrier Bureau of
the FCC   www.fcc.gov/ccb.html

Computer Professionals for   jasper.ora.com/andyo/cyber-rights/
Social Responsibility (CPSR)   telecom.html

Electronic Frontier Foundation   www.eff.org/

Electronic Policy Network   epn.org/
(EPN), Research and Advocacy
Organizations and Publications
Providing Online Analysis of
Economics, Politics, Social Trends,
and Public Policy

Electronic Privacy Information Center	www.epic.org
European Commission Docs (EARN)	www.earn.net/EC/bangemann.html
Federal Communications Commission (FCC)	www.fcc.gov/
Federal Communications	www.law.indiana.edu/fclj/fclj.htm
Ralph Nader and the Consumer Project on Technology	www.essential.org/cpt/telecom/telecom.html
U.S. Long-Distance Pricing	www.teleworth.com

WIRELESS TECHNOLOGY

The Bluetooth Web Site	www.bluetooth.com
Cellular Digital Packet Data (CDPD) Presentation	www.path.berkeley.edu/~guptar/wireless/CDPD/index.htm
The Cellular IP Project at Columbia University: Cellular Wireless Internet Access	comet.ctr.columbia.edu/cellularip/
IEEE 802.11 Working Group (Wireless LANs)	grouper.ieee.org/groups/802/11/
The Infrared Data Association Homepage	www.irda.org
Intel	www.gsmdata.com
The International Wearable Computing Web Site	www.wearcomp.org
Local Multipoint Distribution Service (LMDS)	www.ajs2.com
Personal Communications Service (PCS) Info (Intel)	www.pcsdata.com
Portable Computing and Communications Association (PCAA)	www.pcaa.org
Primer on Wireless WANs by Peter Rysavy	www.networkcomputing.com/netdesign/wireless1.html

(Continues)

University of Waterloo GSM Page	ccnga.uwaterloo.ca/~jscouria/GSM/ index.html
Wireless Application Protocol (WAP) Forum Home Page	www.wapforum.org
Wireless Data Forum	www.wirelessdata.org
wow-com, The World of Wireless	www.wow-com.com/consumer

COMPANY CONVERGENCE

NETWORK HARDWARE AND SOFTWARE MANUFACTURERS

3Com	www.3com.com
ADC Kentrox	www.kentrox.com
Bay Networks	www.baynetworks.com
Cascade	www.casc.com
Cellware Broadband Technologies	www.cellware.de
Cisco Systems Web Page	www.cisco.com
Data Communications' 1999 Global Enterprise Networking Directory	www.data.com/directory/
FastComm	www.fastcomm.com
Fore Systems	www.fore.com
Fujitsu	www.fujitsu.com
General Datacomm	www.gdc.com
Hewlett-Packard	www.hp.com
IBM Networking (Raleigh) Web Server	www.raleigh.ibm.com
Livingston Communications	www.livingston.com
Motorola	www.mot.com
Motorola ISDN Systems Group	www.motorola.com/MIMS/ISG/
Network Systems Corp. (Fibre Channel, HIPPI, 802.6 ftp Server)	ftp://nsco.network.com

Newbridge Networks	www.newbridge.com
NorTel	www.nortel.com
Novell	www.novell.com
Racal	www.racal.com
TranSwitch	www.txc.com
U.S. Robotics	www.usr.com
Wandel & Goltermann	www.wg.com/wg

TELECOMMUNICATIONS CARRIERS

AirTouch PCS	airpcs.com/airpcsHome.html
Ameritech	www.ameritech.com
AT&T	www.att.com
Bell Atlantic	www.ba.com
Bell Canada	www.bell.ca
BellSouth	www.bellsouth.com
British Broadcasting Company (BBC)	www.bbcnc.org.uk
British Telecom Scotland (U.K.)	nsa.bt.co.uk/nsa.html
Cellular One	www.elpress.com/cellone/cellone.html
Finland Telecom	www.tele.fi/
France Telecom	www.francetelecom.com
Frontier Communications	www.frontiercorp.com/
GTE	www.gte.com
Helsinki Telephone Company (Finland)	www.hpy.fi/
IXC	www.ixc-comm.com
KDD Laboratories (Japan)	www.kddlabs.co.jp/
MCI	www.mci.com

(Continues)

Nippon Telephone and Telegraph (Japan)	www.ntt.co.jp/index.html
Pacific Bell	www.pacbell.com
Qwest	www.qwest.net
SBC (Formerly Southwestern Bell)	www.sbc.com
Southern New England Telephone (SNET)	www.snet.com
Sprint	www.sprint.com
Stentor (Canada)	www.stentor.ca
Tampere Telephone Company (Finland)	www.tpo.fi/
Telstra (Australia)	www.telstra.com.au
US West	www.uswest.com
WorldCom/WilTel	www.wcom.com/home.shtm

SERVICES CONVERGENCE

Amazon	www.amazon.com
AMR	www.amr.com
AOL	www.aol.com
Ariba	www.ariba.com
Brio	www.brio.com
Cisco	www.cisco.com
Clarify	www.clarify.com
Dell	www.dell.com
EBay	www.ebay.com
E*Trade	www.etrade.com
Forrester Research	www.forrester.com
Hewlett-Packard	www.hp.com
Hill Associates	www.hill.com
INS	www.ins.com

KPMG	www.kpmg.com
Luna	www.luna.com
Oracle	www.oracle.com
PeopleSoft	www.peoplesoft.com
Prodigy	www.prodigy.com
RightPoint	www.rightpoint.com
SAS	www.sas.com
Siebel	www.siebel.com
Sterling	www.sterling.com

E-COMMERCE-RELATED SITES

AUCTION SITES

Andy's Garage Sale	www.andysgarage.com
AuctionBot (University of Michigan)	auction.eecs.umich.edu
Auction Supersite	www.onsale.com Onsale.com
Auction Universe	www.auctions.com/
Avant Garde Virtual Marketplace	avantgarde.4w.com/
Classifieds 2000	www.classifieds2000.com/
eBAY	www.ebay.com
First Auction	www.firstauction.com
Webauction	www.webauction.com
Yahoo! Auctions	auctions.yahoo.com

AUTOMOBILES

Auto-by-Tel	www.autobytel.com
AutoWeb Interactive	www.autoweb.com

(Continues)

CarPoint (Microsoft)	carpoint.msn.com
Online Auto	www.onlineauto.com

BANKS (ONLINE SERVICES)

Bank America	www.bankamerica.com
b@nk: Atlanta Internet Bank	www.atlantabank.com
ComuBank (Member FDIC)	www.compubank.com
E-Loan	www.eloan.com
Interloan.com	www.interloan.com
iQualify	www.iqualify.com
NextCard Visa	www.nextcard.com
Security First Network Bank	www.sfnb.com
TeleBank Online	**www.telebankonline.com**
Wells Fargo Bank	www.wellsfargo.com

BOOKS

Amazon.com	www.amazon.com
Bargain Book Warehouse	store.bargainbookwarehouse.com/ cgi-bin/sodacreek.storefront
Barnes & Noble	www.barnesandnoble.com
Borders	www.borders.com
O'Reilly and Associates	www.ora.com

CLOTHING, ACCESSORIES, AND HOUSEHOLD

The Bombay Co.	www.bombayco.com
Eddie Bauer	www.eddiebauer.com
Fortunoff Jewelry	www.fortunoff.com
The Gap	www.gap.com
Homewarehouse	www.homewarehouse.com

JCPenney — www.jcpenney.com

Lands' End — www.landsend.com

Macy's E-ssentials — www.macys.com

Pacific Coast Feather Co. — www.pacificcoast.com

Victoria's Secret — www.victoriassecret.com

COMPUTERS AND CONSUMER ELECTRONICS

Beyond.com — www.beyond.com

Buy.com — www.buy.com

BuyComp — www.buycomp.com

BuyDirect.com — www.buydirect.com

BuySoftware.com — www.buysoftware.com

Chumbo.com — www.chumbo.com

CompUSA — www.compusa.com

Computer Discount Warehouse (CDW) — www.cdw.com

Computers.com — www.computers.com

computershopper.com — www.computershopper.com

Cyberian Outpost — www.cyberianoutpost.com

Data Comm Warehouse — www.warehouse.com

Dell Computer — www.dell.com

Egghead Software — www.egghead.com

Internet Shopping Network — www.isn.com

KillerApp — www.killerapp.com

Macromedia — www.macromedia.com

The Mac Zone — www.maczone.com

Microsoft Corp. — www.microsoft.com

NECX — www.necx.com

PC Connection — www.pcconnection.com

(Continues)

The PC Zone	www.pczone.com
Price Watch	www.pricewatch.com/
RoboShopper	www.roboshopper.com
Shopper.com	www.shopper.com

FOOD

Chef's Catalog	www.chefscatalog.com
Cook's Garden	www.cooksgarden.com
Flying Noodle	www.flyingnoodle.com
Godiva Chocolatier	www.godiva.com
Kosher Grocer	www.koshergrocer.com
NetGrocer	www.netgrocer.com
Pepper Plant Hot Sauce	www.pepperplant.com
Pizza Hut	www.pizzahut.com
Virtual Vineyards	www.virtualvin.com
Wine.com	www.wine.com
Wilderness Coffee House	www.wilderness-coffee.com

INSURANCE

Aetna Insurance	www.aetna.com

MOVIES

Movielink	www.777film.com
Online Video Store	www.spcc.com/video/video.htm
Reel.com	www.reel.com

MUSIC

Amazon.com	www.amazon.com

CDNow	www.cdnow.com
CD Universe	www.cduniverse.com
Music Boulevard	www.musicblvd.com
Rock.com	www.rock.com
Tower Records	www.towerrecords.com

NEWS AND SPORTS

CNN Interactive	www.cnn.com
ESPN Sportszone	www.sportzone.com
MSNBC	www.msnbc.com
Nando Times	www.nando.net
National Public Radio	www.npr.org
San Jose Mercury News	www.sjmercury.com
Up-To-The-Minute	www.uttm.com
USA Today	www.usatoday.com
Wall Street Journal Interactive	www.wsj.com
Wired Magazine	www.wired.com/news/

OFFICE SUPPLIES

| OfficeMax Online | www.officemax.com |
| Staples | www.staples.com |

PETS

Aardvark Pet	www.aardvarkpet.com
Petopia	www.petopia.com
Pets.com	www.pets.com

(Continues)

SHIPPING SERVICES

Federal Express	www.fedex.com
United Parcel Service	www.ups.com
United States Postal Service	www.usps.gov

SHOPPING BOTS

Excite shopping	www.excite.com/shopping
mySimon	www.mysimon.com
Yahoo! Shopping	shopping.yahoo.com

STOCK MARKET

E*Trade	www.etrade.com
Investor (Microsoft)	investor.msn.com
Quote.com	www.quote.com

SUPERSTORES

The Catalog Site	www.catalogsite.com
CompareNet	www.comparenet.com
CyberShop	www.cybershop.com
NetMarket	www.netmarket.com
Shop4.com	www.shop4.com
Shopping.com	www.shopping.com

TRAVEL SERVICES

Biztravel.com	www.biztravel.com
Expedia (Microsoft)	expedia.msn.com
Preview Travel	www.previewtravel.com
PriceLine	www.priceline.com

Travelocity	www.travelocity.com
TravelWeb	www.travelweb.com

MISCELLANEOUS

Buy It Online	www.buyitonline.com
Commercepark	www.commercepark.com/
The Internet Mall	www.internetmall.com
The Microsoft Plaza (formerly eShop)	plaza.msn.com/msnlink/index.asp

ONLINE PAYMENT MECHANISMS

ActivMedia	www.activmedia.com
Actra Business Systems	www.actracorp.com
Apache	www.apache.org
AT&T EasyCommerce and SecureBuy	www.att.com/easycommerce/about.html
Baltimore Technologies Ltd.	www.baltimore.ie
BBN	www.bbn.com/offerings
BBNPlanet	www.bbn.com/planet
Beenz	www.beenz.com
Broadvision	www.broadvision.com
C2Net	www.c2.net
CertCo	www.certco.com
CheckFree	www.checkfree.com
Clickshare	www.clickshare.com
CommerceNet	www.commerce.net
ComputerWorld's E-commerce Papers	www.computerworld.com/emmerce/index.html

(*Continues*)

Connect, Inc.	www.connectinc.com
Cryptography FAQ (RSA)	www.rsa.com/rsalabs/faq/home.html
CyberAtlas: The Reference Desk for Web Marketing	www.cyberatlas.com
CyberCash	www.cybercash.com
CyberCharge	www.cybercharge.com
Data Interchange Standards Association (DISA)/ASC X12	www.disa.org
DES Information	www.quadralay.com/Crypto/source-books.html
DigiCash	www.digicash.com
Digital Equipment	www.digital.com
Digital Signature Information from the Software Industry Coalition	www.softwareindustry.org/issues/1digsig.html
EC World Online	ecworld.utexas.edu/
EDIBANX, Bank Alliance Network Exchange (Financial EDI)	www.edibanx.com
Electronic Banking Resource Center at Ohio State University	www2.cob.ohio-state.edu/~richards/banking.htm
Electronic Commerce World	www.ecomworld.com
Electronic Commerce World Institute	www.ecworld.org
Electronic Payments Forum	www.epf.net
Encommerce Inc.	www.encommerce.com
Entrust Technologies	www.entrust.com
Firefly	www.firefly.com
First Virtual Holdings	www.firstvirtual.com
Forman Interactive Corp. (Internet Creator E-Commerce Server)	www.internetcreator.com
Gradient Technologies	www.gradient.com

GTE CyberTrust	www.bbn.com/products/security/cytrust/index2.htm
Harbinger	www.harbinger.com
HipHip Software (Merchandizer Server)	www.merchandizer.com
IBM (HomePage Creator for E-business Server)	www.ibm.com
iCat (Commerce Online Server)	www.icat.com
ICentral	www.icentral.com
ICVERIFY	www.icverify.com
Integrion Financial Network	www.integrion.com
The Internet Economy Indicators	www.internetindicators.com
The Internet Factory	www.ifact.com
Intershop Communications	www.intershop.com
Ironside Technologies	www.ironside.com
JEPI information	www.w3.org/ECommerce/Overview-JEPI.html
Lotus Development Corp.	www.lotus.com
MasterCard	www.mastercard.com
Microsoft	www.microsoft.com
Millicent (Digital)	www.millicent.digital.com
MiniPay	www.ibm.net.il/ibm_il/int-lab/mpay/
Moai Technologies' LiveExchange Auction Software	www.moai.com/
Mondex	www.mondex.com
NetBill	www.netbill.com
NetCash	www.teleport.com/~netcash
NetCheque	nii-server.isi.edu/info/NetCheque
NetChex	www.netchex.com
Netcom (Small Business Pack Server)	www.netcom.com

(Continues)

Netscape	`www.netscape.com`
NetVentures (ShopBuilder Server)	`www.shopbuilder.com`
NIST "Framework for Global Electronic Commerce"	`www.iitf.nist.gov/eleccomm/glo_comm.htm`
Novell	`www.novell.com`
OnLineCHECK	`www.onlinecheck.com`
Open Market	`www.openmarket.com`
Oracle Corp.	`www.oracle.com`
PayMe (W3C)	`www.w3.org/pub/Conferences/WWW4/Papers/228/`
Peachtree Software	`www.peachtree.com`
Pitney Bowes	`www.pb.com`
Premenos	`www.premenos.com`
Pretty Good Privacy	`www.nai.com/products/security/security.asp`
RSA Data Labs	`www.rsa.com`
San Diego Business	`www.businessite.com/Entry/BusinessNet/bj/bjinwells.html`
Security Dynamics Technologies	`www.securitydynamics.com`
S-HTTP Information	`www.eit.com/creations/s-http`
Simple Network Communications (Commerce Made Simple Server)	`www.simplenet.com`
Social Security Administration	`www.ssa.gov`
StarNine Technologies	`www.starnine.com`
Sterling Commerce	`www.stercomm.com`
SunCommerce (SecureMerchant Server)	`www.suncommerce.com`
Symantec	`www.symantec.com`
Terisa Systems	`www.terisa.com`
TradeWave	`www.tradewave.com`
TRUSTe	`www.truste.org`

USWeb	www.usweb.com
VeriFone	www.verifone.com
VeriSign	www.verisign.com
Viaweb	www.viaweb.com
Virtual Spin (Internet Store Server)	www.virtualspin.com
Visa	www.visa.com
The Vision Factory	www.thevisionfactory.com
WebMate	www.webmate.com
Xcert Software	www.xcert.com
Yahoo! (Yahoo! Store server)	store.yahoo.com
ZipLock	www.portsoft.com

INDUSTRY SUPPORT ORGANIZATIONS

2600 Magazine	www.2600.com
Alliance for Telecommunications Industry Solutions (ATIS)	www.atis.org
American Bankers Association	www.aba.com
American Bar Association	www.abanet.org/homeie.htm
American Health Information Management Association	www.ahima.org
American Medical Informatics Association	www.amia.org
American Petroleum Institute	www.api.org
American Society for Industrial Security	www.asisonline.org
Chemical Industry Data eXchange, Inc.	www.cidx.org
Electronic Commerce Canada	www.ecc.ca
Electronic Commerce Council of Canada	www.eccc.org

(Continues)

Electronic Frontier Foundation	www.eff.org
Electronic Funds Transfer Association	www.efta.org
Electronics Industry Data Exchange, Inc.	www.eidx.org
Georgia Tech's WWW Survey	www.cc.gatech.edu/gvu/user_surveys
Healthcare EDI Coalition	www.hedic.org
Information Systems Security Association	www.uhsa.uh.edu/issa
Information Technology Industry Council	www.itic.org
International Organization for Standardization (ISO)	www.iso.ch
International Telecommunication Union (ITU-T)	www.itu.int/ITU-T/index.html
Internet Engineering Task Force (IETF)	www.ietf.org
Internet Society (ISOC)	www.isoc.org
MIT's Media Lab	www.media.mit.edu
Mortgage Bankers Association of America (MBAA)	www.mbaa.org
National Automated Clearing House Association (NACHA)	www.nacha.org
Network Wizard's Internet Domain Survey	www.nw.com/zone/WWW/top.html
Nielsen Media Research	www.nielsenmedia.com
The Organization for Economic Cooperation and Development (OECD)	www.oecd.org
Pan American EDIFACT Board	www.disa.org/paeb
Phrack Magazine	www.phrack.com
RosettaNet	www.rosettanet.org
Smart Card Forum	www.smartcrd.com
Treasury Management Association	www.tma-net.org

Treasury Management
Association of Canada
`www.tmac.ca`

Uniform Code Council (UCC)
`www.uc-council.org`

Vanderbilt University's
Project 2000
`www2000.ogsm.vanderbilt.edu`

Workgroup for EDI
`www.wedi.org`

World Wide Web
Consortium (W3C)
`www.w3.org`

INDEX

AUTHOR BIOGRAPHY

Steven Shepard is a professional writer and educator in Williston, Vermont. He has written several books and many articles on a wide variety of both technical and non-technical topics. He is the author of *Managing Cross-Cultural Transition: A Handbook for Corporations, Employees, and Their Families* (Aletheia Publications, New York), and *A Spanish-English English-Spanish Dictionary of Telecommunications Terms* (Shepard Communications Group).

Mr. Shepard specializes in international issues in telecommunications, the use of alternative delivery media, the development of multilingual educational materials, and the social implications of technological change. He has written and directed more than 40 videos and films and written technical courses on a broad range of topics. He routinely teaches and writes in both English and Spanish and has been published in both languages.

Mr. Shepard received his B.A. in Spanish from the University of California at Berkeley and his M.S. in International Business from St. Mary's College. He is married and has two children.

He can be reached via e-mail at Steve@shepardcomm.com.